TOO
MUCH OF
A GOOD
THING

TOO MUCH OF A GOOD THING

How Four Key
Survival Traits Are Now Killing Us

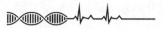

LEE GOLDMAN, MD

Little, Brown and Company
New York Boston London

Little, Brown and Company
Hachette Book Group
1290 Avenue of the Americas, New York, NY 10104
littlebrown.com

First Edition: December 2015

Little, Brown and Company is a division of Hachette Book Group, Inc. The Little, Brown name and logo are trademarks of Hachette Book Group, Inc.

The publisher is not responsible for websites (or their content) that are not owned by the publisher.

The Hachette Speakers Bureau provides a wide range of authors for speaking events. To find out more, go to hachettespeakersbureau.com or call (866) 376-6591.

ISBN 978-0-316-23681-2
Library of Congress Control Number 2015937505

10 9 8 7 6 5 4 3 2 1

RRD-C

Printed in the United States of America

For Jill

Contents

CONTENTS

TOO
MUCH OF
A GOOD
THING

Introduction

Ever since I began practicing medicine, my family and friends have inevitably asked me about real and perceived health problems. Common questions always include: "Why is it so hard for me to lose weight?" "Do I really need to take my blood pressure medicine even if I feel okay?" "Should I take an aspirin every day?"

In thinking about these questions and others, I began to appreciate what seems to be a peculiar conundrum. Some of the key protective strategies our bodies have used to assure the survival of our human species for tens of thousands of years now cause many of the major diseases of modern industrialized societies.

In writing this book, my first goal is to emphasize the historical fragility of the human species and why we wouldn't be here today, let alone dominate the world, if it were not for fundamental survival traits such as hunger, thirst, fear, and our blood's ability to clot. My second goal is to explain why these hardwired survival traits are often now "too good"—not only more powerful than we need them to be to survive in the modern world but also so strong that, paradoxically, they've become major causes of human disease and death. My third and perhaps most important goal is to explain how the future may play out, including how we can continue to use our brains to influence that outcome.

Too Much of a Good Thing focuses on the four key human survival traits, without which we wouldn't be here today:

- *Appetite and the imperative for calories.* Early humans avoided starvation by being able to gorge themselves whenever food was available. Now that same tendency to eat more than our bodies really need explains why 35 percent of Americans are obese and have an increased risk of developing diabetes, heart disease, and even cancer.
- *Our need for water and salt.* Our ancestors continually faced the possibility of fatal dehydration, especially if they exercised and sweated, so their bodies had to crave and conserve both water and salt. Today, many Americans consume far more salt than they need, and this excess salt combined with the same internal hormones that conserve salt and water are the reasons why 30 percent of us have high blood pressure — significantly increasing our risks of heart disease, stroke, and kidney failure.
- *Knowing when to fight, when to flee, and when to be submissive.* In prehistoric societies, up to 25 percent of deaths were caused by violence, so it was critical to be hypervigilant, always worrying about potentially getting killed. But as the world got safer, violence declined. Suicide is now much more common in the United States than murder and fatal animal attacks. Why? Our hypervigilance, fears, and worrying contribute to a growing epidemic of anxiety, depression, and post-traumatic stress — and the suicides that can result.
- *The ability to form blood clots so we won't bleed to death.* Because of their considerable risk of bleeding from trauma and childbirth, early humans needed to be able to clot quickly and efficiently. Now, with the advent of everything from bandages to blood transfusions, blood clots are more likely to kill us than excessive bleeding. Most heart attacks and

strokes—the leading causes of death in today's society—are a direct result of blood clots that block the flow of arterial blood to our hearts and brains. And long car rides and plane trips, unknown to our distant ancestors, can cause dangerous and sometimes fatal clots in our veins.

Well into the nineteenth century, each of these four traits specifically helped our ancestors survive as they tried to avoid starvation, dehydration, violence, and bleeding—which have been among the leading causes of death throughout human history. But now, amazingly, these same four traits collectively explain more than 40 percent of deaths in the United States, including four of the eight leading causes of mortality, and are directly responsible for more than six times the number of deaths they prevent. How could the very same attributes that helped humans not only survive but also dominate the earth now be so counterproductive?

This paradox is the essence of *Too Much of a Good Thing.* For more than 200,000 years and perhaps 10,000 generations, the world of our ancestors changed very gradually. Our genes, which define who we are, also evolved more or less on a parallel path, so our ancestors could adapt and thrive. Then, barely 200 years ago, human brainpower began to change our world dramatically, as the Industrial Revolution signaled the beginning of an entirely new era of transportation, electricity, supermarkets, and medical care. The good news is that life expectancy, which was little different in the early 1800s from what it was tens of thousands of years ago, has just about doubled since then, including more than a six-year gain just since 1990. But the bad news is that our bodies have had only about 10 generations to try to adapt to this new world. Our genes simply can't change that fast, and, as a result, our bodies are lagging behind our environment. Instead of dying from challenges against which our bodies were designed to protect us, we're now more likely to die from the protective traits themselves.

So what will happen next? There are three major possibilities.

One is that everything gets worse—more obesity, diabetes, high blood pressure, anxiety, depression, suicides, heart attacks, and strokes—until those of us with these diseases don't live long enough to have children. Although this possibility may seem far-fetched, we already see obese children becoming diabetic adolescents who will be less likely to have children—let alone healthy children—than their nondiabetic counterparts. But somehow, we should be more than smart enough to avoid this doomsday scenario.

A second possibility is that we collectively will devote more time and effort to being healthier. We'll all eat better, exercise more, and embrace other virtuous lifestyle changes. Unfortunately, these self-help approaches, though sometimes successful on an individual basis, are notoriously unsuccessful across populations. The term "yo-yo dieting"—a reference to the fact that most people who lose weight usually gain it right back—is just one example of the tendency for many short-term successes to be counterbalanced by long-term failures.

The third possibility is that we'll take advantage of modern science—not in isolation but as a key complement to continued attempts to improve our lifestyles. High blood pressure requires medication; stomach surgery is the most successful way to deal with refractory extreme obesity; the chemical imbalances that cause depression often respond best to antidepressant medications; and an aspirin a day really does make sense for some of us. Ongoing scientific advances also hold great promise for future medications that could control our appetites for food and salt or that could safely reset our clotting systems. And with the decoding of the human genome, we're entering an era in which the specific genetic causes of modern diseases may be treated with medicines that target only the responsible gene. These advances augur a new era of personalized precision medicine: treatment specifically designed to address each individual's needs. This growing reliance on medications or even surgery shouldn't be dismissed as moral weakness but rather recognized as the sometimes necessary way to do what we can't do on our own—because our genes simply aren't built that way and can't change fast enough.

Too Much of a Good Thing addresses all these possible outcomes and potential solutions to our present predicament. Each of the first five chapters begins with a story that helps frame the modern situation, and then explains how our current health challenges are the direct result of historically successful human attributes. The final three chapters provide a blueprint for how we can and must continue to use our brains to get our genes and our bodies back into sync with the environment we've created.

PART I

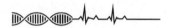

THE PROS AND CONS OF FOUR KEY HUMAN SURVIVAL TRAITS

CHAPTER I

How Our Bodies Became
What They Are

Timothy Ray Brown, who was originally labeled the "Berlin patient" by doctors who tried to protect his anonymity, was in his twenties when he became infected with the human immuno-deficiency virus (HIV)—the virus that causes AIDS—in the 1990s. Doctors successfully treated him with the usual HIV medications of that era, and he did well until he developed acute leukemia. His leukemia was initially treated with aggressive chemotherapy, the side effects of which required that he temporarily stop the anti-HIV medicines. Without these drugs, his HIV blood levels soared, an indication of persistent infection and susceptibility to full-blown AIDS. His doctors had no choice but to reduce his chemotherapy and restart the anti-HIV drugs. The only remaining option for treating his leukemia was a bone marrow transplant, which had about a fifty-fifty chance of curing it. Fortunately, his doctors found a compatible bone marrow donor, and Brown beat the odds—the bone marrow transplant cured his leukemia, and he remains leukemia-free more than five years later.

But the story doesn't end there. His doctors also knew, from studies of unusual people who were infected with HIV but never progressed to severe infection or full-blown AIDS, that just one mutation in a single gene, if inherited from both parents, can prevent HIV from entering cells that it otherwise has no trouble infecting. Unfortunately, less than 1 percent of northern Europeans and even fewer of the rest of us have this fully protective mutation. Remarkably, Brown's doctors found a compatible bone marrow donor who also had this unusual mutation. And the successful transplantation of those bone marrow cells didn't just cure Brown's leukemia — it also made him the world's first person ever to be fully cured of advanced HIV infection.

The first case of AIDS was reported in San Francisco in 1980. What followed was a series of remarkable scientific feats: HIV was confirmed to be the cause of AIDS by 1984; tests to screen for the virus were available by 1985; the first medication to treat it was formulated by 1987; and effective multidrug therapy began to turn HIV infection into a commonly treatable chronic disease by the mid-1990s.

Scientists also looked back and documented HIV in human blood as early as 1959, including in a small number of people who had died of undiagnosed AIDS in the decades before the first reported case. Based on genetic studies of the virus itself, scientists postulate that what we now call HIV originated as a slightly different virus in monkeys in central Africa and probably was first transmitted to humans sometime in the 1920s.

But let's imagine a different scenario: What would have happened if HIV had spread from monkeys to humans before the late-twentieth century advances in virology and pharmacology? HIV infection likely would have spread gradually but persistently throughout the world, especially because infected people are highly contagious for months or years before they develop symptoms. Of course, before the era of modern transportation, the spread initially would have been slower, but we still would have expected the virus to spread wherever humans had intimate contact with one another. If everybody infected with HIV subsequently got AIDS and died, the only remaining humans would either have been those rare people who had this protective mutation or people who were somehow sufficiently isolated or far enough from central Africa to avoid exposure.

The reality, though, is that nothing—no famine, infectious disease, or environmental catastrophe—has ever wiped out the human race. To understand how we've survived, we need to understand where we came from and how we got here. We'll start by going back in time to look at our ancestors.

THE HUMAN FAMILY'S ANCESTRY.COM

My parents always told me that they named me after my maternal great-grandfather Louis Cramer. But when my wife decided to trace both our families on Ancestry.com, census records showed that his name really was Leib Chramoi. The Internet, the most transformative human invention of my lifetime, shed new light on my genealogy and emphasized how much the world has changed in just three generations.

But how about our collective genealogy over thousands of generations? How could our ancestors have survived the myriad threats that antedated HIV but that potentially could have wiped out 99 percent or more of us long before modern medications were available? And going back even further, why did we survive while most species didn't?

If we really want to understand how we got from the first anatomically modern humans—known as *Homo sapiens,* based on our genus (*Homo*) and our species (*sapiens*)—to my great-grandfather Leib, we'll have to understand how our genes define who we are.

And in telling that story, we'll also learn how those same attributes are now a mixed blessing—sometimes still saving our lives but increasingly causing modern maladies.

The earliest archaeological evidence for *Homo sapiens* begins about 200,000 years ago in Africa. Some adventurous *Homo sapiens* apparently first moved out of Africa and into the Arabian Peninsula about 90,000 years ago before dying or retreating back to Africa during the Ice Age because of the cold climate and advancing glaciers. But another group of *Homo sapiens* tried again around 60,000 years ago, migrating first to Eurasia, then spreading to Australasia about 50,000 years ago and later to the Americas about 15,000 years ago.

Archaeological remains also tell us that we weren't the only *Homo* species ever to walk the earth. Based on the dating of fossils, paleontologists estimate that the earliest member of the *Homo* genus, called *Homo habilis,* dates back to around 2.3 million years ago. By 1.8 million years ago, *Homo ergaster* in Africa, then *Homo erectus* in Africa and Eurasia, were as tall as modern humans, although with somewhat smaller brains. Between 800,000 and 500,000 years ago, other *Homo* species in Africa, Europe, and China had skull sizes and presumably brains that were approximately as big as ours. Some of their descendants, including Neanderthals and Denisovans, appeared about 350,000 years ago, followed later by the first *Homo sapiens* about 150,000 years later.

Although paleontologists used to think that each successive *Homo* species replaced the prior, "less human" one and that each more modern species inherited the earth from its predecessor, the reality isn't so simple. Neanderthals and humans, for example, probably overlapped for more than 150,000 years, and we know from fossil remains that humans, such as the cave-painting Cro-Magnons in France and Spain, lived near or even with Neanderthals beginning about 50,000 years ago and continuing for perhaps

10,000 years. Genetic evidence suggests that at least some inter-breeding occurred pretty quickly and that Neanderthal-derived genes now comprise about 2 percent of the DNA of modern Euro-peans and Asians. And since we don't all have the same Neander-thal genes, it's clear that the interbreeding wasn't a one-time phenomenon. We don't know the functions of most of these Neanderthal genes, but as we'll show in chapter 2, they seem to have helped our ancestors adapt better to their new, out-of-Africa environments.

The Denisovans probably lived in Siberia as recently as 50,000 years ago. Based on DNA recovered from a Denisovan girl's tiny pinkie finger bone, scientists conclude that Denisovans interbred with Neanderthals as well as with humans. For example, today's Tibetans, Australian aboriginals, and Melanesians have some Den-isovan DNA. By comparison, Africans have substantially less Neanderthal DNA and no detectable Denisovan DNA — but about 2 percent of their DNA apparently comes from yet another unknown prehistoric *Homo* group that lived in Africa about 35,000 years ago.

Scientists estimate that the Neanderthals across Europe and Asia never exceeded a population of about 70,000 and more likely peaked at about 25,000. And the Denisovan population likely was no larger. As the glaciers receded and the world warmed up, humans eventually outcompeted the brawnier but slower-footed Neanderthals as well as the Denisovans to become the only sur-viving *Homo* species by around 40,000 years ago.

On the one hand, it seems easy to assume that humans sur-vived simply because we're smarter than all other animals. But on the other hand, it's pretty remarkable. It's true that our brains are big, but dinosaurs once dominated the earth despite their propor-tionally tiny brains. To put things into perspective, let's look at it another way: What percentage of species that ever inhabited the earth is still in existence today? The answer: about one out of 500 — just two-tenths of 1 percent.

Yes, we're here today because of the ultimate triumph of our brainpower, but our survival has always been contingent on our brawn—not so much our raw muscular strength but rather all the inherent traits that helped our ancestors survive and even flourish. Our brain's capacity for art, science, philosophy, and technology never could have been realized if our bodies hadn't been vigorous enough to survive in challenging and even hostile environments. Our ancestors had to be able to gather nuts and berries, hunt and kill wild animals that were often bigger and faster than they were, and survive the Ice Age, not to mention droughts and infectious plagues.

In the digital age, a nonathletic, antisocial nerd with poor eyesight and bad asthma can become a billionaire. But for most of human existence, we required robust physical attributes just to survive, not to mention compete for the mate who would help pass our genes on to future generations.

We survived against these daunting odds because of the attributes and resilience of the human species—what we're made of and how our bodies have evolved and adapted over the course of about 200,000 years and 10,000 generations. We're here because no prehistoric predator, climate disaster, or HIV-like infection ever wiped out our ancestors.

NATURAL SELECTION: SURVIVAL OF THE "FITTEST"

Year in and year out, men and women probably always competed for mates mostly in terms of their physical attributes—not unlike the prom king and queen or the football captain and the head cheerleader. But survival of the fittest doesn't mean the survival of the biggest, fastest, or prettiest. Although these physical traits help, they're just the beginning of the story.

We're here rather than the Neanderthals, the Denisovans, or the never-born offspring of humans who failed to procreate because our own ancestors were fitter—they had the wherewithal to meet the earth's environmental challenges and to out-compete others for food, water, and mates. Survival of the fittest explains how natural selection works: over the course of generations, people with the "fittest" genetic attributes live long enough to have the most children, who themselves survive to have more children, and so on. Simply put, if you and your descendants survive and procreate, your DNA will be perpetuated. And if you don't, your DNA, just like your family name, won't.

Our DNA resides on 23 pairs of chromosomes — one member of each pair inherited from each parent. In 22 of these pairs, each of the two paired chromosomes has similar — but not absolutely identical — DNA. The 23rd pair defines our sex — girls inherit an X chromosome from each parent, whereas boys get an X chromosome from their mothers and a corresponding (but smaller and very different) Y chromosome from their fathers.

Each of our chromosomes is composed of a varying amount of DNA, which in aggregate comprise about 6.4 billion pairs of nucleotides, with about 3.2 billion pairs coming from each parent. These 6.4 billion pairs, which are commonly called *base pairs,* can be thought of as the letters that make up the words (genes) that define who we are.

Our chromosomal DNA collectively contains about 21,000 genes that code for corresponding RNA, which then codes for each specific protein that is critical for our bodies to function normally. Surprisingly, these 21,000 genes comprise only about 2 percent of our chromosomal DNA. Another 75 percent or so of our DNA contains about 18,000 more — and generally larger — genes that don't tell RNA to code for proteins but instead send signals that activate, suppress, or otherwise regulate the protein-coding genes or the functions of their proteins. The remainder of our DNA, about 20 percent, doesn't code for any currently known proteins or signals and is sometimes called junk DNA, although scientists may well discover important roles for it in the future.

Since we have two sets of chromosomes, we normally might

expect to have two identical versions of each of our 21,000 genes, one inherited from each parent. But although each half of a child's genes — and his or her DNA — should be an exact replica of the DNA of the parent from whom it came, even the best-programmed computer can make a rare typographical error. When DNA replicates, an imperfection — or *mutation* — occurs only once every 100 million or so times. As a result, each child's DNA has about 65 base pairs that differ from those of her or his parents. Over time, these mutations add up. All humans are about 99.6–99.9 percent identical to one another, but that still means we each have at least six million base pairs that can differ from other humans. These mutations define us as individuals and also show how our human species has evolved over generations.

Mutations themselves are random events. But their fate — whether they spread, persist, or disappear — in future generations is determined by whether they're good (advantageous), bad (disadvantageous), or indifferent (neutral). In natural selection, an initially random genetic mutation is perpetuated if it's sufficiently beneficial to you and your children. By comparison, disadvantageous mutations are usually eliminated quickly from the population even if the child survives. Slowly but surely — across 10,000 or so generations of humans — our genome has changed substantially because of the random appearance but selective perpetuation and spread of beneficial mutations.

Neutral mutations don't spread as widely as beneficial mutations, but they can hang around and accumulate over time. Some neutral mutations are totally unimportant and invisible — our DNA code changes, but we still make the identical protein. It's the same principle as if we were to spell a single word two different ways — such as *gray* and *grey* or *disk* and *disc* — without any alteration in meaning. Some neutral mutations change us — determining, for example, whether a person's chin is dimpled or not — but have no survival impact. But some currently neutral mutations might theoretically become advantageous or disadvantageous in the future. Imagine, for example, a mutation that could make us more or less sensitive to high-dose radioactivity, which isn't a problem now but would become critical if there were a thermonuclear war.

Every now and then, a new advantageous trait might be so *absolutely* critical for the ability to survive an otherwise fatal challenge—such as an epidemic of infectious disease—that it will become the new normal almost immediately, because everyone without it will die. But most beneficial genetic changes simply confer a *relative* competitive advantage, perhaps only in limited geographical locations or in specific situations.

Not surprisingly, the rate at which a genetic mutation spreads depends on how much relative advantage it brings. The spread will be slow at first and then speed up as it disperses more broadly within a closed population. It's like when you bet one chip to win one chip, bet two chips and win two more chips, and keep on going. Once something "goes viral," it increases rapidly because every previous carrier becomes a transmitter—whether we're talking about infectious diseases or a favorable mutation.

Imagine, for example, a mutation that carries a 25 percent survival advantage for you and your children if you get it from both parents and half that advantage if you get it from one parent. This mutation will spread to less than 5 percent of the population in about 50 generations (circa 1,000 years) but to more than 90 percent of the population in 100 generations (2,000 years) and to essentially the entire population by 150 generations (3,000 years). By comparison, if a single new mutation carries just a 1 percent advantage, it will spread more slowly—but it will still spread to nearly 100 percent of a population of a million people in about 3,000 generations, or around 60,000 years. Interestingly, these calculations don't depend much on the size of the population. For example, it will only take about 1.5 times as long for a mutation to spread throughout an entire population of 100 million people as it would take for it to spread to a population of 10,000 people. Slowly but steadily and inexorably, natural selection has defined which genes — and, as a result, which people — inhabit the earth.

NATURAL SELECTION IN ACTION

If every one of us shares a particular gene, it's hard to know what less advantageous gene it may have replaced. But we can get a good appreciation for natural selection by looking at genes that are common or even ubiquitous in some parts of the world, where they presumably have a strong advantage, and rare or even nonexistent in other parts of the world, where they have no advantage or even are deleterious. Two examples are (1) worldwide variations of skin color and their link to our need for strong bones and (2) lactose tolerance and its link to the domestication of animals.

Strong Bones

Adult humans have 206 discrete bones, somewhat fewer than our children, because some bones fuse together as we grow. In aggregate, our bones make up about 13 percent of our body weight. In order to have healthy bones, we need to have their key building block in our diets: calcium. We also need an activated form of vitamin D to help us absorb calcium in our intestines and to regulate its incorporation into our bones.

Without strong bones, we can't grow. In children, vitamin D deficiency causes rickets — a debilitating condition that makes procreation unlikely or impossible because of weak bones, stunted growth, reduced fitness, and childhood death. Although active vitamin D is naturally present in fatty fish such as tuna, salmon, and mackerel as well as in animal livers, kidneys, and egg yolks, historically it's been very hard to get enough active vitamin D from the foods we eat. We can now get active vitamin D in fortified milk, orange juice, and breakfast cereals, but throughout history, our ancestors' bodies actually manufactured much of their own vitamin D. That's because our livers make an inactive precursor to vitamin D that, when activated by ultraviolet light absorbed through the skin, returns to the liver, goes to our kidneys for yet more enhanced activation, and meets our vitamin D needs.

As we'll see in chapter 2, Paleolithic hunter-gatherers typically ate enough animal meat and bones to get their needed calcium, and they were exposed to more than enough sunshine to activate vitamin D. Before humans migrated out of Africa, all our ancestors had dark skin, with its melanin pigment that partially blocks ultraviolet light. But they still got enough ultraviolet light because they spent so much time in the warm sun without needing to shield their bodies with clothes. And when they first moved out of Africa to less sunny climates, dark skin wouldn't have been a problem if they still could get enough sun exposure to activate the vitamin D precursor their livers made or if a hunter-gatherer existence could somehow provide sufficient fatty fish or animal livers or kidneys.

Instead, two things changed. First, when our ancestors migrated out of Africa to colder climates and had to wear more clothes, the amount of skin exposed to sunlight fell dramatically and compromised their ability to activate enough of their livers' vitamin D. Second, when they adopted agriculture about 10,000 years ago, their diet emphasized carbohydrates, their vitamin D intake declined, and they became more dependent on the sunlight to which they then were less exposed. How did they adapt? Random genetic mutations, which reduced their skin's production of melanin and lightened their skin color, spread rapidly because of the advantage they conferred—allowing more ultraviolet light to penetrate their sun-exposed skin and activate the vitamin D precursor made in their livers.

Could vitamin D and bone development really be sufficiently important for survival to explain why light skin replaced dark skin in so much of the world? Let's look for some clues over just the past several centuries. In the mid-1600s, rickets was commonplace among the newly urbanized residents in the foggy cities of England—so serious that many of the children couldn't even walk. In women, it also caused pelvic deformities that interfered with normal childbirth. But it wasn't until 1822, when Jedrzej Sniadecki noted that rickets was much more common in city-dwelling Warsaw children than in rural Polish children, that sunlight was considered a potential way to prevent or cure rickets. By 1890, Theobald Palm reported that poor children in urban Europe developed rickets but equally poor or even poorer children in China, Japan, and India did not. Like Sniadecki, he postulated that lack of sunlight was the cause—and his conclusions, like Sniadecki's, were either ignored or rejected by scientific experts of the time. So it wasn't until 1921, when Alfred Hess and Lester Unger improved the rickets of seven white New York City children by increasing their exposure to sunshine, that the medical community became convinced.

Even today, people who have genetic variations that lower their vitamin D levels modestly for their entire lifetimes have a shorter life expectancy than people without those variations. The enormous survival benefit of anything that helps humans get enough vitamin D explains why we now fortify milk, orange juice, and breakfast cereals to help city dwellers who don't get enough sunshine. Sun exposure and skin pigmentation also explain vitamin D deficiency in modern America—about 6 percent of elderly Americans, who get less sun exposure than younger people, and nearly 30 percent of African Americans are still relatively deficient in vitamin D.

The first persistent skin-lightening mutation appeared after humans migrated out of Africa but before modern European and Asian populations dispersed. But what's interesting is that light skin isn't caused by a single mutation or even different mutations in a single gene. Some mutations are found only in Europeans and some only in Asians, and the wide variety of mutations in different genes helps explain the variations in skin color among northern Europeans, southern Europeans, and people in various parts of Asia and the Americas.

Now that we understand why Europeans and Asians have light skin, we must ask a different question. Why was dark skin so important in Africa? And why today do genes that code for dark skin appear to be essentially identical in black-skinned Melanesians and Australian aboriginals as they are in Africans — suggesting that no mutations were required when their ancestors migrated from Africa to their new homes? Surely a random mutation for light skin must have appeared every now and then in sub-Saharan Africa, Melanesia, or Australia. Why didn't it spread, or why didn't at least a few humans with light skin survive and perhaps form local communities, even if as relative outcasts?

Maybe dark skin protected against overheating or having too much activated vitamin D. The first hypothesis doesn't really make any sense, because dark colors tend to absorb heat whereas light colors tend to repel it. Just think about cars — a black car with a dark interior gets extremely hot during the summer, whereas a white car with a light interior stays relatively cool. Too much vitamin D is seen occasionally in people who drink large amounts of artificially fortified milk and is characterized by the deposition of calcium in places we don't need it, such as in kidney stones. But we don't need dark skin to protect us from getting too much vitamin D, because no more than about 10–15 percent of vitamin D gets activated by ultraviolet light, regardless of how often or how intensely skin is exposed.

How about protection from sunburn? It's true that sunburned skin reduces the activity of our sweat glands and makes it harder to dissipate heat — a key survival trait that I'll emphasize in chapter 3. But people with all but the lightest skin are able to tan, a process that protects them from severe sunburn.

Could it be because of skin cancer? People with dark skin are much, much less likely to get serious forms of skin cancer than are people with light skin. Skin cancer is generally a disease of later adulthood, well after our ancestors generally had their children. But, as I'll discuss in chapter 6, some Paleolithic men continued to be productive — both as experienced hunters and as fertile fathers — into their middle-age years, and women's roles as grandmothers undoubtedly increased the likelihood that their grandchildren would survive. However, skin cancer alone isn't nearly common enough or important enough to drive the evolution of skin color.

It turns out that the best explanation for why dark skin is protective in environments with a lot of sun exposure is our need for *folate* (also called *folic acid*), a B vitamin that's critical for human development and health. Inadequate folate in utero leads to severe neurological abnormalities — and that's why pregnant women are routinely prescribed folate supplements and why bread in the United States and many other countries is now fortified with folate.

Ultraviolet light, absorbed through the skin, alters folate from its active form into an inactive form, thereby causing functional folate deficiency. If you're going to be exposed to a lot of sunlight, dark skin is a great way to protect your folate activity and help you and your descendants survive.

In trying to balance the need for vitamin D on the one hand as compared with folate on the other, humans ideally would have dark skin in parts of the world where sun exposure is especially prolonged and intense and light skin in parts of the world where you'd expect less cumulative exposure to the sun's ultraviolet rays. Remarkably — or, I would argue, not so remarkably, given natural selection — that's exactly what we see. Although there are occasional exceptions, the average skin pigmentation of a local, indigenous population and its ability to tan correlates very closely with its natural exposure to ultraviolet light, which generally is well correlated with a population's distance from the equator.

Skin color is, however, a bit darker in the Southern Hemisphere than at the same latitude in the Northern Hemisphere because ultraviolet radiation levels, corrected for latitude, are actually a little higher in the Southern Hemisphere.

The importance of skin color for human survival wasn't just about becoming lighter as humans moved into Europe and Asia. Skin color could also become dark again, and that's exactly what happened when, for example, light-skinned northern Indians migrated farther south on the Indian subcontinent.

The one notable exception—which, I will argue, proves the rule—is the Inuit people, whose dark skin color seemingly contradicts the vitamin D hypothesis. But the traditional Inuit diet is chock-full of fatty fish and also includes the occasional animal liver, so sunlight was never critical for meeting their vitamin D needs. And although there isn't a huge downside to ultraviolet light in the Arctic, reflection of sunlight off the ice (which increases a person's exposure to ultraviolet light by as much as 90 percent) presumably was enough of a stimulus for the natural selection of somewhat dark skin.

The advantages of dark skin in very sunny areas and light skin in less sunny and cooler areas isn't a uniquely *Homo sapiens* phenomenon. The powerful effect of latitude on skin color was also evident in Neanderthals. The genome of European Neanderthals indicates that they had light skin—a clearly beneficial mutation, especially for a species that spent a lot of time in dark caves. But interestingly, the genetic mutation that lightened the skin of European Neanderthals was different from any of the mutations that explain the light skin in modern Europeans. And although the cave-dwelling Neanderthals of southern Europe had light skin, Neanderthals in sunny climates, such as the Middle East, had dark skin. Pigmentation may be only skin deep, but the mutations that affect it were critical for human—and even Neanderthal—survival.

Lactose Tolerance

Another striking example of natural selection relates to our ability to digest milk. Mammals are defined by the presence of mammary glands that make the milk their offspring need to survive. Since all infant mammals initially depend entirely on their mothers' milk for nutrition and hydration, the ability to digest that milk is an absolute survival requirement—but only until the child is old enough to live off of other foods.

Milk is a nutritious combination of water, fat, proteins, calcium, salt, and sugar. Milk's sugar, called *lactose,* is indigestible until it's broken down in the small intestine into its constituent parts—one molecule of glucose and one molecule of galactose, which are easily absorbed into the bloodstream. This breakdown of lactose requires an enzyme, *lactase,* which is critical for the survival of infant mammals. Like all human enzymes, lactase is a protein, which in this case is coded by a gene on chromosome 2. Close to the gene that codes for lactase is another gene that codes for the protein that activates the lactase gene.

As exemplified by two of our favorite pets—dogs and cats—all mammals naturally lose their lactase and, as a result, their ability to digest lactose. After weaning, dogs should never be given milk, which in large quantities can make them desperately ill. And although we often see cute pictures of kittens drinking milk, it turns out that most adult cats are also lactose intolerant and develop the same sorts of symptoms that plague lactose-intolerant humans. For most of human existence, all our ancestors also lost essentially all their lactase activity and became lactose intolerant by the time they reached approximately five to seven years of age.

Why should mammals become lactose intolerant? For one thing, a mother's lactation suppresses her ovulation and ability to conceive again, especially if she is relatively undernourished. If she kept on breast-feeding, she would have far fewer children. And when it becomes time for the next child to suckle at her breast, the new infant's survival shouldn't be jeopardized by having to fight an older sibling over a limited supply of milk. So it makes perfect sense for young mammals gradually to lose their lactase activity as they grow and can obtain and eat other foods.

Without lactase, lactose passes undigested through the small intestine and into the large intestine, where it draws in water like a sponge as part of a process that equalizes water concentration on each side of the permeable intestinal membrane. This extra water in the intestine causes the cramps and diarrhea often associated with lactose intolerance. But it gets even worse. The lactose in the large intestine is great food for bacteria, which have no trouble breaking it down into glucose and galactose and using these sugars for their own nutrition. In the process, hydrogen gas, methane, and carbon dioxide are produced, and much of this gas is then expelled as flatus. Although the percentage of undigested lactose (which produces diarrhea) as compared with bacterially digested lactose (which produces gas) may vary among lactose-intolerant individuals, neither is a desirable outcome. Of course, the degree of symptoms depends not only on whether someone is totally lactose intolerant but also on how much milk he or she drinks.

So let's be clear. Lactose intolerance isn't abnormal, and lactose-intolerant people are not deficient mutants. Just the opposite! All human adults were lactose intolerant for about 190,000 years, or perhaps 9,500 generations of human existence, until our ancestors first domesticated cattle and subsequently goats and camels. This domestication began in Egypt about 9,000 years ago, in the Middle East about 8,000 years ago, and in sub-Saharan Africa about 4,500 years ago. And with these domesticated animals came the opportunity to use their milk as well as their meat for human nutrition.

So perhaps it's not surprising that something dramatic happened about 7,000 years ago, within 2,000 years of the first domestication of cattle: a random change appeared in just one base pair of the lactase-activating gene. And just one copy of this mutation was enough to keep the lactase gene permanently in the "on" position.

Over the past several thousand years or so, this ability to digest the lactose in fresh milk provided a substantial survival advantage that's been estimated to be about 4–10 percent higher than it is in people who can't digest lactose. Although milk may not be "nature's most perfect food," and although many of us are perfectly healthy even though we never "got milk" after early childhood, substantial data document the benefit of milk and milk products for bone growth and human height, particularly when other foods are in short supply. This benefit makes sense because a cow's milk can provide needed water, salt, and calcium. It's also an efficient way to get nutrition — a cow's milk provides five times as many calories per acre of feed as its meat and almost twice as much as its cheese.

The survival advantage of lactose tolerance allowed these new mutant humans to migrate with their livestock and displace hunter-gatherers in many parts of the world. Today, about 95 percent of northern Europeans, 70–85 percent of central Europeans and Americans who emigrated from Europe, 80 percent or more of herders from sub-Saharan Africa, and 70 percent of northern Indians have lactase persistence. By comparison, rates of lactase persistence are only about 10–20 percent among nonherders of sub-Saharan Africa, extremely rare in East Asians and Southeast Asians, and generally no higher than 30–40 percent in most other parts of the world.

But what's especially remarkable is that different mutations occurred in the same lactase-activating gene in various parts of the world where cattle and other milk-producing domesticated animals were prevalent. Although a single mutation explains most lactase persistence in central and northern Europeans, several totally independent mutations explain it in southern Europe and northern Africa, another explains it in northern China, and three others are found among lactose-tolerant Tibetans. Although experts aren't sure exactly which of these mutations appeared first, the ultimate spread of lactase persistence to about 25 percent of the world's population is a classic example of the following three evolutionary principles.

First, a random mutation in the lactase-activating gene could have occurred in any person anywhere in the world, independent of whether animal milk was available in the adult diet. But its perpetuation and spread among that person's descendants with access to fresh milk wasn't random. It was an example of what might be called a *niche benefit* — an advantage that's perpetuated in ecological niches where it's advantageous but likely to disappear in niches where it isn't.

Second, the places where we now see lactase persistence may not necessarily be the places where the genetic mutation originated. For example, lactose tolerance is especially common in Scandinavian countries, where milk consumption is a relatively new phenomenon. The genetic mutation found in Scandinavians likely originated in the Balkans and in central Europe, where the mutation that explains lactase persistence in Indian cattle herders also originated.

Third, the finding of six or more lactase-persistent mutations in different niches is, like skin pigmentation, an example of convergent adaptation or evolution, in which different random mutations in our DNA result in the same physiologic adaptation. It really didn't matter what single base pair changed in the activating gene for lactase as long as it provided the same benefit by keeping the gene active and allowing milk sugar to be broken down into adulthood.

PROTECTION FROM INFECTIONS

Just as humans couldn't survive without strong bones and adequate nutrition, our ancestors also needed to be able to resist a wide range of infectious diseases. For the most recent two generations of humans, by far the most serious new infectious epidemic is HIV.

HIV replicates itself rapidly by entering human cells and hijacking our native resources for its own use. Although occasional untreated patients can remain asymptomatic for 20 or more years, in most HIV-infected people the *viral load,* which is the concentration of HIV virus in their blood, rises to more than 10,000 individual viruses in each milliliter (or cubic centimeter) of blood within several years. These viruses infect and damage many types of cells throughout the patients' bodies. Most critically, HIV enters and destroys T lymphocytes, the very same cells that are crucial for warding off many infections. In the absence of effective medications, people with a high concentration of HIV in their blood routinely lose most of their T lymphocytes, develop full-blown AIDS, and die — either from the direct effects of the virus itself or, more commonly, from other infections that they are no longer able to resist.

Given the genetic diversity among humans, we shouldn't be surprised that the estimated 33 million people infected with HIV progress to AIDS at varying rates. Even before the advent of effective medications, about one HIV-infected person in 500 was called a *long-term nonprogressor* — someone who never develops either the direct complications of HIV itself or the infections that otherwise are rampant in people whose immune systems are destroyed by HIV. In some nonprogressors, blood levels of HIV remain relatively high but are somehow well tolerated. In others, HIV levels remain low or even disappear.

Although some long-term nonprogressors are infected with less virulent forms of HIV, others are blessed with rare genetic mutations that make them innately more resistant to HIV. Of these mutations, probably the most common and best studied is the Delta 32 mutation of the *CCR5* gene, which codes for a protein on the surface of T lymphocytes. To infect our cells, HIV first needs to attach to the cell's surface. In about 85–90 percent of infections, the normal CCR5 protein serves as that receptor. After attaching to CCR5 on the lymphocyte's surface, HIV enters the cell, disables it, and ultimately destroys it. When the Delta 32 mutation prevents HIV from binding to the surface of T lymphocytes, most viral strains can't enter the cell and kill it. Interestingly, a different *CCR5* mutation, Delta 24, reduces CCR5 activity in 98 percent of red-capped mangabeys (a type of monkey) and makes them resistant to the monkey version of HIV.

About 10 percent of northern Europeans, 5 percent of southern Europeans, and a smaller percentage of people from other parts of Europe and India have inherited this mutation from one of their parents, but essentially no one from Africa or other parts of Asia has it. People who have one copy of this gene are about 25 percent less likely to have their HIV infections progress to AIDS. But really lucky people have two copies of this mutation — one from each parent. In these people, the most common forms of HIV can't get into the T cells at all. AIDS doesn't develop, and the body eventually rids itself of all or nearly all of the virus. Since approximately 10 percent of people of northern European descent have one copy of this mutation, simple arithmetic (a 10 percent chance of getting it from one parent × a 10 percent chance of getting it from the other parent) indicates that about 1 percent of that population will get it from both parents, and these two copies will make them completely resistant to most HIV infections.

So when Timothy Ray Brown's doctors were looking for a bone marrow donor to treat and potentially cure his leukemia, they faced real challenges. First, of course, they needed to find a donor whose blood was as compatible as possible with Brown's — this way, the transplant would repopulate his bone marrow without being rejected after chemotherapy had killed not only the leukemia but also his normal bone marrow. But they also looked for and found a donor who had two copies of the *CCR5* Delta 32 mutation, in the hope that the donor's bone marrow cells would make Timothy Ray Brown resistant to HIV.

Amazingly, that's exactly what happened. Brown stopped taking all his HIV medications and has remained free of both leukemia and HIV for more than five years. For a person with leukemia, five years of cancer-free survival is usually considered a cure, and we know that a bone marrow transplant can cure about 50 percent of people with the type of leukemia that Timothy Ray Brown had. But for HIV, we have no precedent — Timothy Ray Brown is the first and, at least for now, only person apparently to be cured of full-blown HIV infection.

Since two copies of the CCR5 Delta 32 mutation apparently provide complete protection and one copy provides partial protection against the progression of most forms of HIV, it's not surprising that scientists would try to develop drugs that could block the unmutated CCR5 receptor. The hope is that such drugs could duplicate the natural protection and be effective for treating HIV and AIDS. In fact, such medications have been developed, are generally well tolerated, and appear to decrease the number of T cells that are destroyed by HIV. In perhaps 50 percent or so of infected people, however, the HIV virus itself mutates and finds an alternative way to attach to a cell's surface and perpetuate the infection. In these patients, as well as the 15–20 percent of patients initially infected by less common forms of HIV, blocking the unmutated CCR5 receptor can't cure the infection. So it's not surprising that an occasional person can be chronically HIV infected despite having two copies of the beneficial CCR5 mutation.

Now that we know that the CCR5 mutation, if inherited from both parents, can often prevent HIV infection, the next question, of course, is, why do some people have it? Did the ancestors of the 10 percent or so of northern Europeans with the mutation from one parent (half a dose) realize some benefit from it? After all, anything that's present in 10 percent of a local population may be there for a reason. And interestingly, this mutation was also present in about 10 percent of Europeans 7,000 years ago, during the Bronze Age.

Scientists initially conjectured that *CCR5* mutations might have protected our ancestors against well-documented infectious epidemics throughout recorded history — such as bubonic plague or smallpox. But studies have shown no evidence that *CCR5* mutations protect against bubonic plague, and any possible protection against smallpox remains speculative.

Perhaps somewhere in pre–Bronze Age Europe, this mutation protected against another infectious disease that no longer exists. Scientists estimate that 300,000 or more different viruses may currently infect various mammals. For example, bats have been implicated as the initial hosts responsible for recent outbreaks of rapidly fatal human Ebola virus and Middle East respiratory syndrome (MERS), which is almost as deadly as the severe acute respiratory syndrome (SARS) outbreak a decade earlier. Rodents are thought to be responsible for rapidly fatal human Lassa fever and hantavirus infections. And chimpanzees were the original source of HIV/AIDS, before it spread to the human population.

The common denominator shared by these viruses and all infectious organisms is that they need to find a host who can harbor them without dying too quickly. Even viruses and bacteria are subject to the laws of natural selection. Over the short term, infectious agents do best if they can infect a lot of animals and replicate as fast as possible in them, regardless of whether they kill everything they infect. But over the long term, the surviving viruses and bacteria need someplace to live, or they'll cease to exist. That seems to be why the SARS virus disappeared — all its hosts were either killed by it or slaughtered as part of public health efforts to eliminate infected animals.

At this point, we really don't know why the *CCR5* Delta 32 mutation first arose or why it has persisted in approximately the same percentage of northern Europeans for 300 or more generations. But we do have some inkling as to why the *CCR5* mutation never took hold in Africa. The *CCR5* mutation appears to increase the likelihood that the West Nile virus will cause more serious disease—and the West Nile virus has historically been common in Africa but rare in Europe. Presumably the Delta 32 mutation was either a neutral or slightly advantageous mutation for northern Europeans, but any potential advantage in Africa was more than offset by its West Nile disadvantages.

At any rate, the presumed selection advantage for the Delta 32 mutation in Europe was realized before the Bronze Age. Then, in an unpredictable twist of fate about 300 generations later, the Delta 32 mutation had a totally unanticipated benefit: it protected against a brand-new infectious disease—HIV—that, ironically, originated in central Africa, where essentially no one has the mutation.

THEN AND NOW

The common occurrence of light skin in nontropical climates—and the clear gradation of lighter skin color the farther you get from the equator—is a classic example of *genetic sweep,* which is an advantage so substantial it's seen in everyone in the population. Although skin pigmentation may be the clearest visual example of genetic sweep, we should think of it not as unique but rather as proof of a principle. If mutations in skin color became almost perfectly correlated with exposure to ultraviolet light within less than 40,000 years, think of all the other mutations that have swept across the human species before or perhaps even since then. Our DNA is replete with mutations that have been equally critical for human survival, many of which long antedated the need to adapt to varying latitudes or levels of sun exposure.

But many beneficial mutations don't carry as big an advantage as skin pigmentation, so they spread less quickly. As a result, our overall genetic makeup has changed rather slowly over the 90,000 generations of evolution since *Homo erectus;* the 10,000 generations since modern *Homo sapiens* first appeared; the 600 or so generations since the advent of agriculture, about 12,000 years ago; the 280 or so generations since the domestication of the horse in Kazakhstan first provided an alternative to human transportation, about 5,600 years ago; and the ten generations or so since the appearance of the first steam locomotive, in 1801, heralded the era of modern mechanized vehicles.

Beyond its slow speed, another major limitation of natural selection is that it can only favor what's best right now — there's no window into the future to predict what mutations may be better years or generations from now. If the world changes rapidly within a generation or over the course of a very small number of generations, the "win-now" benefits of prior natural selection can be totally irrelevant for the new challenge.

For 200,000 years, Homo sapiens has generally flourished because this combination of a slow rate of random genetic change along with a win-now force of natural selection has usually been consistent with an equally slow rate of change in our environment. Of course, every now and then something dramatic might happen — such as the Ice Age, which brought cold weather and caused humans to retreat to warm climates. And periodic war, famine, or infection could threaten survival to the extent that only selected humans, who were better able to adapt or migrate, would live to perpetuate their advantageous adaptations.

But let's compare the typically slow changes in our environment over millennia with the rapid changes since around the beginning of the nineteenth century, when the Industrial Revolution began to alter the world dramatically. New machines, electricity, gasoline-powered vehicles, modern appliances, and computers have transformed our world in no more than ten generations or so. Although droughts and famines still occur in some parts of our planet, in other parts we have turned near deserts, like California's Sacramento Valley, into veritable gardens. The growing food supply has allowed the human population to increase from about one billion in 1800 to 2.5 billion in 1950, six billion in 2000, and seven billion in 2011. And as our nutrition and sanitation have improved, childhood mortality rates have plummeted from the historic worldwide average of about 50 percent, which persisted until as late as 2000 in isolated parts of Mali, to a worldwide average of about 9 percent in 1990 and less than 5 percent in 2013. Even in sub-Saharan Africa, with its poverty and high prevalence of HIV, the childhood mortality rate is only about one-quarter of the old historic average. This reduction in childhood mortality has been the biggest driver of the increase in human life expectancy, which probably never exceeded about 30 years worldwide until the late nineteenth century and is now more than 71 years worldwide and more than 80 years in approximately 30 countries. At this rate, the world's population will exceed eight billion by 2024.

People of all ages now benefit from improved access to food and water, fewer threats and injuries from wild predators, better sanitation, antibiotics, vaccines, and other aspects of modern medical care. As a result, we live in an increasingly aging society. The worldwide median age (meaning that 50 percent of us are older and 50 percent younger than that age) has risen from about 24 in 1950 to nearly 30 in 2010. In the same time frame, the median age in Europe rose from 30 to 40.

In fact, the human condition has improved precisely because we as a species no longer spend all our time hunting and gathering food or even tending to our farms or domesticated animals. The time previously devoted to these simple survival tasks can now be spent on creating science, technology, art, and other advances that define a modern civilized world. We have become a species of gradually aging, sedentary people who live indoors, ride in cars, take elevators, and may or may not engage in occasional exercise. And most of that change has occurred in 200 years—barely the blink of an eye in the 200,000-year history of *Homo sapiens* and the two-million-year history of the *Homo* genus.

Let's now consider the modern conundrum. Our brains have

allowed us to make extraordinary changes in our environment, at speeds and in directions that never could have been predicted. By comparison, our bodies, which were designed to "Xerox" our DNA nearly perfectly in each subsequent generation, continue to evolve in slow motion. Our genes just can't keep up.

And why not? Simple arithmetic. For a genetic mutation to spread from one person in a thousand to nearly 100 percent of the population in about 10 generations—equivalent to the 200 years or so since the onset of the Industrial Revolution—it would need to carry an enormous survival advantage, so big that people with just one copy of it would be more than 30 times more likely to reproduce successfully than people without it. But since more than 98 percent of Americans already live long enough to have children who, in turn, live long enough to have children, no genetic mutation can have enough of a competitive advantage to spread that quickly and that widely in today's population. The only exception would be if we were suddenly faced with an unprecedented survival challenge—for example, if everyone became HIV infected but no effective treatments were available.

TOO MUCH OF A GOOD THING

The traits that allowed our ancestors to survive through food shortages and droughts, to recognize and avoid dangerous situations, and to clot when cut or bruised were critical for the survival of our species. But the very recent imbalance between the rapid rate at which we have changed our environment and the slow rate at which our genes can change explains why we're stuck with genetic traits that were finely tuned over millennia to deal with a pre–industrial age world.

Our genes can't possibly mutate fast enough to keep pace with the rate of change in today's world. And as long as modern killers afflict us *after* we bear children who will in turn have their own children, there's no natural selection process to give an advantage to genes that hypothetically could help us catch up. As a result, the same survival traits that so successfully perpetuated our species are often overly protective and sometimes frankly deleterious in a modern era in which food, salt, and water are available in excess, violence is at an all-time low, and we rarely bleed to death.

To invert a well-known aphorism, we've won the war of human survival, but we're losing a battle with adaptation. Aging adults are developing a whole series of chronic diseases—some simply because we live longer and others because these conditions are the side effects of traits that were once critical to our ancestors' survival. To understand the everyday consequences of this conundrum, we'll begin by talking about food and hunger, then move on to salt and water, memory and fear, and bleeding and clotting.

CHAPTER 2

Hunger, Food, and the Modern Epidemics of Obesity and Diabetes

In 1908, the physician and anthropologist Ales Hrdlicka toured the southwestern United States and detailed, for the first time, the health of the indigenous peoples, including the Pima Indians who lived along the Gila River in Arizona. He saw just one case of diabetes among the 4,000 or so Pimas, and his formal report of 11 common diseases, 11 occasional diseases, and 6 rare diseases never mentioned diabetes. In 1937, the famous diabetologist Elliott Joslin, who founded the eponymous Joslin Diabetes Center in Boston, visited the same reservation and reported that 21 Pimas had diabetes, just about the number he expected based on the rate of diabetes in the overall US population. By 1954, however, the frequency of diabetes among the Pima Indians increased more than tenfold. And by 1971, 50 percent of adult Arizona Pimas had diabetes — the highest rate anywhere in the world.

This extremely high prevalence of diabetes turned out to be the direct result of another condition that was also affecting the Pima population in record numbers. The Pimas had a startling 70 percent prevalence of obesity — the highest rate in the world.

At first, the Pima Indians were thought to be an anthropological curiosity. Perhaps they had somehow developed a unique genetic ability—honed over centuries of unreliable food supplies—to avoid starvation by needing fewer calories than the rest of us. Then when food became plentiful, they were uniquely cursed to become obese.

But now we know that the Pimas aren't as unusual as we may have thought. Rates of obesity and diabetes are soaring worldwide, especially as diets Westernize in developing countries. In the United States, more than one-third of us are obese (meaning we're more than 20 percent above the upper limit of recommended weight), about another one-third are overweight but don't meet formal criteria for being obese, and a remarkable 10 percent of us have diabetes. Some parts of the world have obesity rates that either approach or, in the case of the island nation of Nauru, match or even exceed that of the Pimas, and global diabetes rates are soaring.

How did the Pimas become the initial warning sign of an emerging worldwide epidemic? Unfortunately, it's distressingly simple. Since the origin of our species, humans have always had a powerful desire for the types of foods that provide the calories our bodies require. We can process large volumes of food, so we can gorge when food is plentiful, storing any excess intake as fat in order to survive periodic fasts. And we have the ability to turn a variety of foods into the energy we need. Since starvation could kill not only an individual but also potentially the entire species, all our instincts and internal thermostats were set to err on the side of wanting too much food and being able to absorb more than what's required to meet our immediate needs.

The human body was never programmed to expect a chronic excess of available food, especially food that could be obtained without burning large numbers of calories just to hunt or gather it. So as steady supplies of food became widely available, problems like obesity and diabetes started to spread.

THE SURVIVAL CHALLENGE

The human body requires energy to fuel all its needs. Energy is measured in calories: one calorie is the amount of energy needed to increase the temperature of one kilogram of water by one degree centigrade. We get almost all our calories from three main sources: fat, carbohydrates, and protein. Fat has about nine calories per gram, whereas carbohydrates and protein each have about four calories per gram. We also get about seven calories per gram from pure alcohol, about three calories per gram from organic acids such as the citrate in citrus fruits, and even fewer calories from artificial sweeteners and digestible dietary fibers. So if we're just worried about calories, eating fat is by far the most efficient way to meet our needs.

The energy our bodies need at total rest—just to keep us alive and not even kicking—is called our *basal metabolic rate*. Basal metabolism varies widely among us because, all things being equal, it goes down as we age but up if we're tall and especially if we're heavy. And for a given weight, it's a little higher if we're muscular and lower if we have a lot of fat. But even after accounting for these known factors, basal metabolism can vary by about 20–25 percent among otherwise similar people—and some of these variations tend to run in families for reasons we often don't understand.

Approximate basal metabolic needs for women can be calculated as follows:

$$(4.5 \times \textit{weight in pounds}) + (16 \times \textit{height in inches}) - (5 \times \textit{age in years}) - 161$$

And for men, the general formula is:

$$(4.5 \times \textit{weight in pounds}) + (16 \times \textit{height in inches}) - (5 \times \textit{age in years}) + 5$$

For the sake of simplicity, let's assume you're forty years old, of approximately average height, and just at the top of the normal range of weight for your height. What we'll get is an average basal metabolic rate of about 1,300 calories in women and about 1,700 calories in men. Most of those calories are used to keep our organs functioning—the two biggest consumers are our brains and livers—but about 10 percent of the calories are burned like logs in a fireplace to provide the heat needed to maintain our normal body temperature. Then let's add in another 40 percent or so in incremental calories needed for the minimal exertion performed by a modern couch potato for a total of about 1,800 calories per day in women and about 2,400 per day in men.

Interestingly, the levels of activity that we see in modern life don't influence daily caloric needs nearly as much as we might think. To maintain a steady weight, our approximate caloric consumption should be about 1.4 × basal for a sedentary person, 1.6 × basal for somewhat active people, 1.8 × basal for moderately active people, 2.0 × basal for very active people, and up to 2.5 or so × basal for soldiers on active duty and athletes in training. And the exercise that determines these levels is often less than you'd expect. For example, a person who sleeps for eight hours (0.95 × basal), sits for four hours (1.2 × basal), and walks around doing routine non-strenuous activities for 12 hours (2.5 × basal) would be classified as moderately active and need about 1.8 × basal calories.

Using these estimates, a moderately active, normal-weight, 5-foot-10-inch American man doesn't need more than about 3,100 calories per day—only 700 or so more than if he were a modern couch potato! And a very active, forty-year-old, 5-foot-4-inch, 145-pound American woman would, on average, burn about 2,600 calories per day. But it's important to keep in mind that every number we use for basal metabolism and average caloric expenditure is really an approximation, which could be off by at least several hundred calories in either direction—higher or lower.

BACK TO THE PALEOLITHIC ERA

If we only need about 1,800–2,400 calories a day to maintain our weight as total couch potatoes, how many calories did our hunter-gatherer ancestors need? For clues to the past, let's look at some surviving hunter-gatherer societies. At one extreme are the Tarahumara, who live in an isolated mountain area in northern Mexico and are known for their endurance. They run kickball races over more than 75 miles at an altitude above 7,000 feet. In a full 15–17-hour day of running, they can expend as many as 11,000 calories, as high a calorie-burning rate as ever reported. But of course not every Tarahumara races, and those who race don't race every day. The highest reported *average* caloric needs for existing hunter-gatherers are for the Aché of Paraguay—about 3,600–3,800 per day.

Perhaps a more representative example comes from the Hadza foragers of northern Tanzania, who maintain their traditional lifestyle in a savanna woodland. Women gather plants, while men gather honey and hunt game on foot. The Hadza are substantially more active than industrialized Westerners, but they're also much smaller; the men average only 112 pounds, and the women average only 95 pounds. After adjusting for their weights, the Hadza's basal metabolic rates are similar to those of modern Westerners. Overall, Hadza men expend an average of about 2,600 calories per day and Hadza women an average of about 1,900 calories per day.

Based on these examples, let's estimate that our male hunter-gatherer ancestors burned an average of about 3,000 calories per day, while the women burned about 2,300 calories per day, depending on their size. But regardless of their exact caloric needs, our ancestors had to be able to eat a lot when food was available in order to protect themselves when it wasn't. How much was a lot? One of our best estimates comes from the surprisingly meticulous food diaries of Meriwether Lewis and William Clark, who were sent by president Thomas Jefferson to explore the Louisiana Purchase in the early 1800s. At that time, the native tribes that populated the American West were among the last and most successful of the world's large-scale, predominantly hunter-gatherer societies. Lewis and Clark's party, who patterned their eating habits after these tribes, consumed 12,000 or more calories per person per day—up to nine pounds of meat each—on days when game was plentiful!

Of course, Lewis and Clark didn't really burn that many calories every day. Let's look, for example, at modern versions of strenuous or prolonged physical exertion. If we hike for six hours a day, half of it while carrying a heavy backpack—the modern equivalent of carrying back the food a hunter killed or a gatherer gathered—we burn about 450 calories per hour and perhaps 6,000 calories over the course of a day. Modern ultra-athletes who run 45 miles per day burn an estimated 6,300 calories. Cyclists can burn up to 1,000 calories per hour in intense racing and as many as 8,000 calories per day, still well short of the Tarahumara extreme of 11,000 calories per day. The Lewis and Clark party didn't need 12,000 calories on any particular day. Instead they were doing what humans have done since we first set foot on this earth: gorging during times of plenty to tide themselves over for times when food was hard to get, as it often was on their journey.

In 1985, Boyd Eaton and Melvin Konner of Emory University in Atlanta described how Paleolithic nutrition differed from a modern diet. The Paleolithic diet emphasized wild meat, fish, nuts, roots, vegetables, and honey. And of course it didn't include farm-raised meat, grains, the milk or milk products of domesticated animals, refined sugars, processed oils, or alcohol. Eaton and Konner estimated that, on average, our ancestors ate more protein, calcium, and fiber; much less saturated fat and salt; and about the same amount of cholesterol as we do.

But for small bands of omnivorous hunter-gatherers, there never was a single Paleolithic diet. Unlike carnivorous lions, hoofed herbivores, or the fastidious koala bear—which lives off the leaves of eucalyptus and gum trees—humans can and will eat almost anything that isn't spoiled or poisonous, so we aren't limited to ecological niches that can provide specific types of food. The principal exception is that we can't digest most grasses and leaves, whose starches are composed of cellulose, which passes as undigested fiber through our intestines.

Of course, we don't know exactly what our early ancestors ate each day. Much of our information comes from surviving hunter-gatherer societies that remain isolated in locations where there's enough food to survive. These localized societies probably don't represent hunter-gatherers who spread to all four corners of the earth when they exhausted the local food supply or grew too large in numbers to subsist on it. And if the food supply itself migrated— as bison do when they search for greener pastures—then human hunter-gatherers would follow.

In all surviving hunter-gatherer cultures, animals are always part of the food supply. According to the research of Loren Cordain and his colleagues at Colorado State University, the 230 or so existing hunter-gatherer societies get, on average, about 60 percent of their calories from animals and about 40 percent from plants. And whereas no surviving hunter-gatherer societies get more than 85 percent of their calories from plants, 20 percent are almost entirely dependent on fished or hunted animals.

Toward the vegetarian extreme are the !Kung bushmen of the Kalahari Desert, who only get about one-third of their calories from meat and the remainder from plants, especially the mongongo nut, which is rich in fat and also has substantial protein. The Hadza of Tanzania rely on honey as well as plants and meats. Some Californian hunter-gatherers were critically dependent on acorns. In the Papua New Guinea rain forest, foragers depend on the carbohydrates of wild sago palms, but the men also hunt and fish for nearly 100 different animal species, with some bringing in more than two pounds of edible food per day.

More carnivorous were the pre-Columbian native Americans of the midwestern plains, who, because of plentiful bison—literally protein on the hoof—were one of the most successful nomadic hunter-gatherer societies in human history. Nevertheless, berries and other plants rounded out their diet when meat was in short supply.

But the real carnivorous extreme is exemplified by Alaskan Inuits, who get nearly all their 3,000 daily calories from animals, including fish. And when we talk about eating animals, we mean the entire animal! Their diet is about half fat, one-third protein, and the remainder carbohydrate—not from plants but rather from the glycogen stored in the muscles and livers of the animals they eat. Glycogen, which is starch composed of a core protein surrounded by as many as 30,000 units of glucose, is a high-energy animal source of nearly pure sugar—the simplest form of carbohydrate.

The best estimates are that hunter-gatherer diets were, on average, about 20–35 percent fat, but predominantly unsaturated fats, from nuts and lean, grass-fed wild game—unlike the saturated fats we currently get from overfed livestock animals. As a result, our ancestors' cholesterol levels were undoubtedly very low, just like those of surviving hunter-gatherer societies. The well-known modern Atkins diet, which also relies heavily on animal protein, gives a far less desirable mix of fats, unless you eat wild game and seal blubber rather than marbleized steaks and bacon.

Hunter-gatherers likely had highly variable carbohydrate intakes, ranging from 35 to 65 percent of their diets. But the sources of carbs varied widely from place to place. Carbs might come predominantly from fruits and berries, from carb-rich tubers, from gathered grains and vegetables, or sometimes only from animal glycogen. And, to prove that we are what we eat, some of this variability can be documented by analyzing the variations in the type of carbon isotopes that were incorporated into the enamel of our ancestors' teeth.

Beyond Calories: Other Nutrients Our Bodies Need

In addition to energy from calories, our bodies need at least 13 different vitamins and 17 different minerals as well as 9 different amino acids (the building blocks of proteins) that we can't make ourselves. A dietary deficiency in any of these needed substances can cause a specific disease. For example, as noted in chapter 1, we need calcium and vitamin D to make our bones. Another important example is vitamin C. The human body can't make vitamin C—also called ascorbic acid—which is needed for the health of collagen, a key building block of skin, bones, and carti-lage. Vitamin C is plentiful in most fresh fruits, especially citrus fruits, berries, and potatoes. Inadequate intake of vitamin C pre-cipitates scurvy, a disease that became prominent in the European seafaring era and probably killed more eighteenth-century British sailors than naval battles did. After decades of controversy, British ships were finally stocked with limes to prevent scurvy, and Brit-ish seamen (termed limeys, not always affectionately) were able to stay healthy even on long voyages.

Interestingly, all mammals other than primates *can* make vita-min C; that's why a lion never needs to eat a fruit salad. And that's why our ancestors, like today's Inuits, did fine if they ate enough whole animal carcasses, whose *fresh* flesh, organs, fat, blood, and bone marrow provided all their needed nutrients, including vita-min C, vitamin A, calcium, and folate. Even the polyunsaturated fatty acids that are critical for children's developing brains came from eating animal brains.

By comparison, no carnivore can live by eating pure protein alone. In fact, we probably couldn't survive if more than about 40 percent of our calories came from protein. Although our livers can convert fat and, to a lesser extent, protein into the sugars our bodies need, there is a limit to their ability to do so. Inuits have big livers to help make the sugars that come mostly from the fat of the fish and polar game they eat, not from the carbohydrates that are missing from their diets. But hunters who rely solely on very lean meat will, paradoxically, wither away despite ample protein because their livers simply can't turn the protein into all the other things they need.

Perhaps the most famous demonstration of the safety and even healthfulness of the whole-animal/animal-only diet was under-taken by the Arctic explorer Vilhjalmur Stefansson, who had observed that the Inuit could live for nine months on nothing but animals, including fish. In 1928, he and his colleague Karsten Anderson agreed to an experiment. In a hospital under close supervision, they ate nothing but beef, lamb, veal, pork, and chicken, includ-ing the animals' muscles, livers, kidneys, brains, bone marrow, bacon, and fat. After following the diet in the hospital for about a month, they resumed their regular activities but documented their adherence to the diet with a variety of biochemical tests for a full year. The only exception was that Stefansson occasionally ate a few eggs and a little butter when he was on trips and couldn't obtain meat. The only carbohydrate in their diet was the animals' stored glycogen, which amounted to no more than 2 percent of the 2,000–3,100 calories they ate per day.

Although they each lost several pounds, Stefansson and Ander-son remained remarkably healthy, with no deficiencies in any vita-mins or minerals and no other abnormalities that doctors could detect back then. As noted by Stefansson, eating the food only lightly cooked was probably important for providing the necessary vitamin C, because vitamin C gets partially oxidized (think of iron turning to rust) and hence deactivated by high temperatures.

A vegan diet is more challenging, because no single fruit, veg-etable, or starch can provide the full range of nutrients that our body needs. That's not a big problem for modern vegans, who have access to a cornucopia of options — but it was a challenge for pre-historic foragers, who didn't have access to animal meat, milk, fish, or eggs. We can now get a lot of protein from the beans and some grains that proliferated after the development of agriculture, but these options weren't common in the Paleolithic era, and there isn't much protein in fruits and berries. That's why nuts and even insects were often key components of hunter-gatherer diets.

HOW WE'RE BUILT TO MEET THE CHALLENGE

Because malnutrition and starvation have been perpetual threats to human survival, it's not surprising that we're built to desire food — especially some of the key foods we need — and to protect ourselves from being sickened or killed by contaminated or poisonous food. Our bodies rely on a number of hormones and organs to drive and control hunger, taste, and digestion. After all, we're the descendants of people whose genes helped them eat and digest enough calories and store enough fat to survive periodic food shortages and to perpetuate the species.

Hunger and Satiety

The brain, and specifically the hypothalamus — which is located behind the nose toward the center of the brain — serves as "command central" in regulating both hunger and satiety (the feeling of fullness). To do so, it continuously interprets incoming data about whether more food is needed and then emits appropriate messages, some of which encourage eating while others tell us not to eat. At least 20 of our body's natural molecules and hormones are involved in this complicated cross talk.

Some signals come from the gastrointestinal system. For example, there's only one intestinal hormone, ghrelin, that tells us to eat — and if we have it in excess, for whatever reason, we stay hungry and just keep on eating. But when we eat, the stomach, intestines, pancreas, and gallbladder typically release a variety of hormones to tell the brain that our food needs are being addressed. Just the distention of the stomach sends back a message that we've had enough food.

Signals also come from fat tissue, which releases the hormone leptin to tell us to eat less. Low leptin levels stimulate us to eat more and to try to burn fewer calories. Leptin-deficient individuals are constantly hungry—and although they collectively represent only a very small percentage of obese people, they develop marked obesity, which can be reversed by giving them leptin. In rare cases, leptin levels themselves are normal, but the brain is unable to sense leptin's presence; such individuals are very obese and don't respond to being given yet more leptin, which their brains can't detect.

Taste

The hypothalamus may tell us when we're hungry, and feedback from our intestinal tract helps us know when we're full. But what we eat isn't driven by either the hypothalamus or the intestinal tract; it's driven by our sense of taste. And the sense of taste begins with our taste buds.

We have thousands of taste buds along the anterior surface of the tongue, in the grooves toward the back of the tongue, on the sides of the tongue, and on the roof of the mouth. The location of the taste buds shows why licking something with the tip of your tongue isn't the best way to appreciate its full taste. Rather, a wine taster's seemingly ostentatious approach—swishing the wine around so that it comes into contact with the entire tongue as well as with the roof of the mouth—actually makes sense and isn't just a way of showing off.

Each taste bud has about 50–150 individual taste cells, each of which specifically detects one of the five tastes: sweet, umami, salt, bitter, and sour. Some think the sensation of fatty taste may be a sixth taste sense; but even if it is, the sweet sense seems to be more powerful.

Each taste sensor in each taste bud sends an individual signal to the brain whenever it senses its specified substance. And, remarkably, each specific type of taste sensor, regardless of where it's located, sends its message to a particular place in the brain—the sweet center, the umami center, the salty center, the bitter center, or the sour center.

Sweet and umami tastes are considered uniformly positive, and if a food has these tastes, it tastes good. Umami—think steak with Worcestershire sauce—senses certain amino acids as being savory. In humans, umami is stimulated most strongly by L-glutamate, which, when combined with sodium, becomes a salt—monosodium glutamate (MSG). That's why MSG is used as an additive to enhance taste.

Salty taste is also desirable, except when the salt concentration is very high. Our preference for salt and salty foods helps us avoid dehydration (as we'll review in more detail in chapter 3). But there's a limit to our love of salt—we wouldn't want to drink ocean water, whose very high salt content could be toxic to our cells. Interestingly, the reason we don't like food or drinks with very high salt concentrations is that they activate both the bitter and sour taste receptors, and this protective phenomenon dissuades us from eating really salty food.

Bitter taste is unpleasant—a protective mechanism because many poisonous plants are bitter. Of all the taste cells, the bitter cells are the most sensitive, able to detect even very small amounts of bitter substances, presumably because it's always been important to avoid even very small amounts of poison. We actually have more than 25 different genes that help us sense bitter tastes, compared with a total of just three genes to sense both sweet and umami.

The ability to sense bitter tastes varies among us. Back in the 1930s, DuPont chemist Arthur Fox accidentally released a cloud of phenylthiocarbamide (PTC) in his laboratory, which didn't bother him at all but led a lab colleague to complain about a very bitter taste. Subsequent testing showed that about 70 percent of us can taste PTC but the rest of us can't. The percentages vary among different ethnic groups, and perhaps it isn't surprising that those of us who can taste PTC are less likely to like things that taste bitter: we're less likely to smoke cigarettes, drink coffee, and enjoy plants like turnips, cauliflower, broccoli, and Brussels sprouts. The same sensitivity to bitter taste explains why some of us don't like saccharin-sweetened beverages — saccharin tastes sweet at low concentrations but bitter at high concentrations.

Sour taste, which humans generally find unappetizing, is conveyed by acids, which are common in spoiled food contaminated by bacteria. This dislike for sour food is nearly ubiquitous throughout the animal kingdom. But an interesting exception is the fruit fly, which lives off rotting fruit and actually finds the sour taste highly desirable.

The general aversion to sour taste also influences how much we like carbonated sodas. In addition to the sour taste of carbonic acid, which is formed when carbon dioxide is dissolved in water under high pressure, an enzyme on the surface of our sour-sensing cells interacts directly with the carbon dioxide that's released to form the soda's bubbles. The double-barreled sour signals explain why some of us don't enjoy carbonated beverages.

Our tongues can also sense a number of things other than these five tastes. For example, we can sense cold, either as a true temperature or as the coolness of menthol, peppermint, or spearmint. And we can sense heat, either temperature-related or from spicy foods such as chili peppers, wasabi, horseradish, and hot mustard. Tannins in tea and alcohol give an astringent sensation.

What we think of as *flavor* is an incompletely understood integration of taste, aroma, texture, and other sensory inputs. For example, we like foods that smell good — the wine taster, for instance, starts by sniffing the wine. And we like sweets even more when they're combined with fat, which gives them a more pleasant texture.

But other influences besides taste and flavor also affect our food preferences. For instance, we can develop an acceptance or even desire for things that taste bitter or sour — an *acquired taste*. Few people initially like the taste of coffee, beer, or lemons, but many of us eventually like those tastes. The process of acquiring a taste seems to be independent of any biological changes in our taste buds. Instead, it apparently is related to social cues — knowing that something we're eating or drinking isn't going to make us sick despite its taste — or its other desirable effects, such as the stimulation of coffee or the relaxation of alcohol. But other than variations in bitter taste, we don't really know very much about why some of us like some foods and others like other foods.

We do know, though, that our taste buds explain some interesting past and present dietary habits across the globe. Our desire for sweetness explains not only modern affinities for sugar-containing foods and beverages but also the attraction to honey, which was an important staple of many of our hunter-gatherer ancestors and remains a major source of calories for the Hadza foragers of Tanzania.

L-glutamate, which is present in breast milk, stimulates the umami receptor and helps explain why infants naturally suckle. Other umami-rich substances include soy sauce, fish stock, Parmesan cheese, shiitake mushrooms, and ripe tomatoes. If you like meat, you probably like these foods, too.

Another interesting question is why we get tired of eating only one thing, such as chocolate, even if we love its taste. Although this phenomenon has been termed "palate fatigue," it really isn't driven by our taste buds. Our taste buds and the brain's taste center continue to be happy if we just eat chocolate, but a more influential part of our brain shouts out "Enough!" and tells us to seek more variety in our diet.

The brain's control of appetite encourages us to shift from one desirable food to another. If a variety of desirable foods — each with a different taste, texture, or even appearance — is available, we're programmed to switch around and keep on eating, even to the point of feeling sick. This phenomenon explains why we almost always have room for dessert — a tasty and usually irresistible change of pace — even if we're completely full after a big dinner. From an evolutionary perspective, this urge for diversity helped our ancestors get the variety of nutrients they needed.

An extreme example of how variety drives intake comes from the remarkable experiments of Barbara Kahn and Brian Wansink, who showed that people seek variety in food appearance even if its taste is monotonous. People who were offered M&M's of many different colors ate about 40 percent more than people who were offered the same amount of M&M's but with less color variation — even though M&M's all taste the same regardless of their color.

The relationship between our taste buds and our food needs also explains most of our cravings. We like protein because we need to eat the amino acids our bodies can't make. But since we'll waste away on protein alone, we crave the sugar that will provide the needed carbohydrates. A craving for chocolate or desserts — the proverbial "sweet tooth" — is a good example. And we crave salt, especially when we need it (as I'll describe in more detail in chapter 3). A pregnant woman's desire for pickles is driven by her body's high salt needs.

Our taste buds and our preference for food diversity undoubtedly helped our ancestors balance their diets. But other than our desires for sweet, salty, and umami tastes, there's little evidence that we selectively crave a specific food because it uniquely contains something we need. The only known exception is the phenomenon of pica, in which iron-deficient individuals, especially children, eat clay, paint, ice, dirt, and sand. But even this craving isn't uniformly appropriate, since red clay and paint have iron, but ice certainly doesn't. And vitamin C–deficient British sailors didn't crave any of the foods that could have prevented scurvy. So ultimately, if our desires for sweet, salt, and umami as well as our general preference for dietary diversity don't fulfill our needs for critical nutrients, we can get into big trouble.

Although we like food diversity, we reach a point where we stop eating because the appetite for more food is counterbalanced by the signals that tell us we've eaten enough. But satisfying the desire for the often bland foods our foraging ancestors hunted and gathered was far easier than turning off the desire for today's wide variety of readily available sweet, savory, and perfectly salted modern foods, with smells, textures, and visual appearances that make them even more irresistible. And unfortunately, the historic benefit of shifting from nuts to berries isn't realized when a modern human shifts from a Big Mac to fries.

Modern food choices make it much easier for us to overeat and, as a result, become overweight or even obese. And the same principle helps explain why indigenous populations who are newly overweight after switching to a Western diet — such as obese Australian aborigines and native Hawaiians — rapidly lose weight when put back on their traditional diets, even if given access to unlimited amounts of those traditional but comparatively monotonous foods.

Digestion and Absorption: Turning What We Eat into What Our Body Needs

Eating gets us started on the road to nutrition, but then our bodies have to process the food and turn it into the calories and other key nutrients we need. The normal digestive process begins in the mouth, where food is chewed, moisturized by saliva, and partially digested by some of the enzymes in that saliva. The esophagus serves as little more than a pipe to transport the food to the stomach, where acidification continues the digestive process and also kills many bacteria.

Nutrients in our diet are absorbed almost exclusively in the small intestine, where different nutrients are preferentially absorbed in different sections. For example, iron is absorbed principally in the first part of the small intestine, the duodenum — in about the same location where the gallbladder secretes its bile to aid in subsequent fat absorption and where the pancreas secretes enzymes that aid in the absorption of both fat and protein. Most absorption occurs just a little bit downstream, in the jejunum, but vitamin B_{12} and many fats are absorbed in the ileum, which is the final portion of the small intestine. We can do pretty well with fat absorption even without a gallbladder, but if you lose your pancreatic enzymes you have serious limitations in your ability to absorb proteins and, especially, fat. If parts of the small intestine, which normally measures about 20 feet overall, are progressively removed, the absorption of nutrients can be compromised, although we generally can do okay until less than about six feet of small intestine remains.

The large intestine, which measures about five feet on average, is where we absorb much of the water we drink or that's contained in the food we eat. Our feces are composed of varying amounts of water, undigested and indigestible food, a number of cells shed each day from the internal lining of our intestines, and a bunch of bacteria that come along for the ride. If the pancreas and intestines are working normally and we haven't eaten any food they can't handle — such as lactose in lactose-intolerant people or gluten in people with gluten intolerance — we usually absorb nearly all the calories in the food we eat. The only major outlier is cellulose, whose starch we can't digest and which passes with our feces as fiber.

Our liver functions as the processing plant that turns the things we absorb into the things we need. Essentially all nutrients absorbed by the intestine first go via the portal vein to the liver, which turns them into forms that can be used by the body's other organs or stored as fat. To guide its work, the liver has a variety of thermostats that sense whether our blood has the proper levels of sugar, cholesterol, and other key substances so that it knows whether to make more or less of them. For example, our blood cholesterol levels depend much less on the amount of cholesterol we eat than on the amount of cholesterol made by the liver, which is partly a function of how much saturated fat we eat and partly a function of where the liver's cholesterol-sensing thermostat is set.

Each of these steps—hunger, taste, digestion, absorption, and processing—has been a critical component of providing the energy and other substances required by humans' complex metabolism. Any inadequacies threatened survival of the individual and even the species, so natural selection always preferred genes that performed these functions as robustly as possible.

WHEN OUR SURVIVAL TRAITS ARE INADEQUATE

If we don't get enough calories to meet our body's energy needs, the body has no alternative but to break down our own fat or muscle to provide that energy. After about six hours of fasting— sooner if we're active, a bit longer if we're sedentary or asleep—our body begins to run out of sugar (glucose), which is stored as glycogen in the liver, and it needs to start burning some of our stored fat. If we don't eat for seven days, the brain, which normally runs totally on glucose, switches its metabolism to get about 75 percent of its energy from fat. In modern times, the same metabolic consequence occurs in people, mostly young women, who have anorexia nervosa, a severe eating disorder in which low food intake or induced vomiting reduces body weight below 85 percent of normal.

To protect against starvation, every part of the body slows down except the liver, and our body reduces its baseline caloric needs by 15 percent after a week of starvation and by 25 percent after two weeks. The slowdown in the all-important immune system, which protects against outside invaders, explains why infection is the leading cause of death in starving people.

Perhaps the most relevant data on how long it takes to starve to death come from the 1981 hunger strikes staged by Irish Republican

Army prisoners, who considered themselves political prisoners and not criminals (although some had been convicted of attempted murder and even murder itself). Their protest was designed to pressure prime minister Margaret Thatcher—known as the Iron Lady—into recognizing their special status by exempting them from typical prison rules, such as wearing prison garb and performing prison work. The original and most famous was Bobby Sands, who was elected to the British House of Commons during his hunger strike. Twenty-two other prisoners joined the cause when they sequentially embarked on hunger strikes, during which they consumed nothing but water. Sands and nine of his colleagues, all previously healthy young men, starved themselves to death—the quickest death occurred after 46 days and the slowest death occurred after 73 days. The average time between refusing food and death clustered at a little more than 60 days. Of the 13 survivors, only one held on for more than nine weeks before he and others were taken off the hunger strike by their families.

By the time of their deaths, these ten initially lean prisoners had each lost an average of about 55 pounds, or about 35 percent of their original weights. Of course, the heavier you are to start, the longer you can be starved. On average, starvation is fatal when people lose about 40 percent of their body weight, which usually corresponds to a 30–50 percent loss of body protein and a 70–90 percent loss of body fat. The record for nonfatal starvation is 382 days without any calories by a man who started out at about 450 pounds and lost about 280 of them—a 60 percent weight loss. But he continued to take in fluids, vitamins, and minerals, so he wasn't truly starving.

If healthy young people lying around in a jail cell die from starvation in about 60 days, how long could our ancestors go while actively seeking whatever they could forage? Let's try some quick arithmetic. If a starving prisoner, like a couch potato, needs about 1,800 calories per day and a hunter-gatherer desperately searching for food needed perhaps 2,700 calories per day, a hunter-gatherer who started at the same baseline would perish about one-third faster, or within about 40 days. But even those 40 or so days would be possible only because of all the traits that helped our ancestors eat as much as they could when food was plentiful.

Of course, it's not that simple. Our ancestors probably started off with even less stored fat than the Irish Republican Army prisoners, and their weakening bodies would have been subject to a wide range of threats not found in an isolated jail cell. And, as we'll see in chapter 3, if progressive debilitation also limited access to water, death would occur much more quickly.

THE MARCH OF CIVILIZATION

Agriculture and the domestication of animals, both of which began about 10,000 years ago, markedly changed the human nutritional equation and made it possible for an acre of land to feed more people than it could before. When our ancestors identified cultivatable plants, they focused on rich sources of calories, depending on what they found growing in the wild wherever they lived. Their staples became grains such as wheat, barley, millet, corn, and rice; sugarcane; tubers and roots such as potatoes, sweet potatoes, yams, and taro; peas, beans, and legumes of various sorts; and melons, including squash and eggplant. Herders tended to move around with their livestock — especially sheep, cows, and goats — which turned the grasses that humans can't digest into ready sources of protein and fat. And some settlements, especially as they got bigger, had plants *and* animals — a diversified food supply that, in the most hospitable settings, allowed great civilizations to rise and flourish.

But even then, survival wasn't simple. Studies of modern hunter-gatherers show that they can hunt or gather about ten calories for every calorie expended while hunting and gathering, a 10:1 yield for their efforts. For subsistence farmers, the ratio is slightly higher, about 12:1, but yields can be as high as 50:1 with irrigation and animal assistance, even in the absence of mechanization. However, agricultural societies often must feed people who aren't directly involved in cultivation, so the people who are actually doing the farming commonly need to generate and consume more calories than hunter-gatherers did because their work also must provide the food needed by other people who don't work on farms.

Since agriculturists couldn't grow everything, they became dependent on a limited selection of plants—and sometimes critically dependent on a single dominant plant—as their source of calories. Droughts, soil depletion, blights, or insects could quickly threaten the local source of calories and lead to widespread famine, so the food supply was no more reliable over the long term than it was for many hunter-gatherer societies. And to make matters even more precarious, the increasingly large settlements of agriculturists were far less mobile than a band of roving hunter-gatherers. As agricultural populations grew, beginning about 10,000 years ago, so, too, did the risk of famine.

For example, let's look at Ireland, where, in 1844, newspapers first reported what would be the 25th failure of their potato plants since 1728. But this time it would be worse—much worse. The estimated Irish potato yield fell precipitously, from nearly 15,000 tons in 1844 to a low of 2,000 tons in 1847 because of the potato blight, which was caused by the water mold *Phytophthora infestans,* which came from America, spread throughout Europe, and thrived in the wet, cool Irish climate.

Despite prior periodic crop failures, the Irish population had grown from about four or five million people in 1800 to more than eight million people by 1841, supported principally by the calories provided by the potato—the single best nonanimal source of almost complete human nutrition. A medium-size potato has about 100 calories, mostly from starch. But it also has about three grams of protein. An Irish diet of 25 potatoes per day could provide not only 2,500 calories but also enough protein, B vitamins, iron, and most other nutrient needs, including vitamin C, to keep a person healthy. The only key nutrients missing from potatoes are calcium and sodium. The Irish potato was so prolific and nutritious that a single acre of potatoes and the milk of one cow — which contains the needed calcium and sodium — provided enough sustenance for a farming family. And by some estimates, potatoes may have been the only solid food eaten by about 40 percent of the notoriously poor rural Irish population.

When the Irish potato crop failed, the rural Irish had no alternative food source, and famine rapidly ensued. Although the precise impact of the famine will never be known, the best estimates are that one million Irish died from starvation or starvation-related illnesses and that well over another million emigrated, predominantly to northern cities in America.

The Irish potato famine is only one in a long list of catastrophic famines — mostly caused by droughts, wars, or political turmoil — recorded since the advent of agriculture. In biblical times, Egypt reportedly withstood seven years of famine only because it had stored large amounts of grain during the preceding seven years of plenty. In the 1600s, famine in northwestern China eventually led to the collapse of the Ming dynasty. In nineteenth-century China, four successive famines killed probably as many as 45 million people. In the Chinese famine of 1959–61, the government reported 15 million more deaths than otherwise would have been expected. Famines in India in the 1630s, early 1700s, 1770s, and 1780s each killed more than two million people. About two million people probably died during famines in France in the 1690s. Famines in Russia killed an estimated five million people in 1921, about one million people during World War II, and more than a million people in 1947.

For twenty-first-century humans, just two grains — rice and wheat — account for about 40 percent of all our calories. As demonstrated by the Irish potato famine, a disproportionate dependence on any single local source of calories has always made humans susceptible to starvation and famine. So it's not surprising that crop failures, exacerbated by political strife, have led to twenty-first-century famines in Somalia, Ethiopia, Kenya, Bangladesh, Afghanistan, and Myanmar. Past and recent famines would have killed even more people if not for all the ways our bodies try to protect us from starving to death.

THEN AND NOW

We really don't know much at all about when during the day most hunter-gatherers ate, but the assumption is that, like other foraging animals, they typically ate whenever they found food. With the advent of agriculture and the domestication of animals, a more reliable food supply made mealtimes more predictable. Wealthy Greeks usually had a midday lunch and an evening dinner, which was probably the bigger of the two meals and, for some, the only meal. Roman physicians recommended a small dinner at around 11:00 a.m., followed by a large supper at around 5:00 or 6:00 p.m. By medieval times, three meals a day became the norm, often with the biggest meal at midday. The reemergence of dinner as the largest meal of the day generally coincided with the development of cities and with men working outside the home then returning at the end of the workday. In Britain, afternoon tea became an additional snack between lunch and dinner.

A modern individual's decision to eat a meal is governed as much by external cues — social, cultural, and environmental — as by immediate need, except when the body faces severe energy shortages. Since food is now readily available throughout the day, it's easy to return to our hunter-gatherer roots and eat whenever food is available — which now means all day long! Modern humans can easily overeat whenever signals of satiety don't supersede the allure of the tasty food we can readily obtain.

Paradoxically, as food has become more plentiful, our caloric needs have declined because of mechanized transportation and a decrease in the need for manual labor. For example, societal changes in twentieth-century America resulted in about 140 fewer calories being burned at work each day by men and about 125 fewer calories by women. In Britain, calories burned over and above basal metabolism declined by 65 percent between 1956 and 1990. Because a relatively small number of farmers, ranchers, and workers in the food chain now do all the labor needed to provide our collective caloric needs, most of us don't have to expend any calories to hunt, gather, farm, or deliver our food if we don't want to.

For hunter-gatherer societies, the search for food was the major form of exercise. Now that we generally don't need to go farther than a supermarket or restaurant to get all the food we need, it's not surprising that about 25 percent of the modern US population is almost totally sedentary.

Average Height as a Marker of a Population's Nutritional Success

Although we now think of an optimal diet as one that promotes longevity, historically our optimal diet was whatever made us strong enough to beget healthy children and then live long enough to help those children survive to do the same. And any animal's best measure of good nutrition is its average size. For humans, this means average height, which is about 80 percent genetic but also strongly linked to childhood and adolescent caloric and especially protein intake.

Although big and strong people might outhunt, outgather, and even outfight their smaller neighbors, they also need more calories. So people faced with chronic or recurrent food shortages are better off if they are short and lean. The small bodies of less-well-nourished humans are not only an effect of their lower intake of calories, protein, and calcium but also protection against the likelihood of even lower intakes in the future.

By these metrics, our nomadic hunter-gatherer ancestors were remarkably well nourished. Remains from the Paleolithic Stone Age show that men had an average height of about 5 feet 10½ inches and an estimated weight of about 148 pounds. The women were substantially smaller, averaging an estimated 5 feet 2 inches and 119 pounds. After agriculture began, about 10,000 years ago, in the Neolithic era, humans actually got smaller. Archaeological evidence suggests that the average Neolithic man was about 5 feet 5 inches tall and weighed 139 pounds, while women were an average of 4 feet 11 inches and weighed 99 pounds.

At first glance, it seems incomprehensible that farming and the domestication of animals would lead to smaller body size. Could it have been all genetic selection? For example, the Pygmies of West Africa are notoriously short, about seven inches shorter on average than the neighboring Bantus, despite rather similar diets, because of mutations in genes that affect the part of the pituitary gland that makes growth hormone. And the offspring of two basketball players would certainly be taller than the offspring of two gymnasts. Maybe the bigger Paleolithic guys and gals really were genetically selected because they were the only ones big enough and strong enough to compete with the coexisting Neanderthals. And maybe by thousands of years later, when the Neanderthals were gone and food was domesticated rather than gathered and hunted, size was no longer as important, so smaller people were more likely to survive and reproduce. But we have no reason to think that human genetics for height dramatically changed between preglacial and agricultural societies, so nutrition—at least in terms of protein—is by far the most likely explanation for why humans got shorter and stayed shorter until very recently.

Paleolithic men were bigger than modern European men through the end of the nineteenth century and were as big as they are today. And when the Europeans came to North America, as supposedly well-nourished invaders from the most technologically advanced countries on earth, the indigenous hunter-gatherer population of the American plains was far taller than they were.

Why would that be the case? It all came down to nutrition. The plains diet of the hunter-gatherers centered around grass-fed wild bison, whose meat was less fatty than meat from modern corn-fed animals. And the plains tribes ate pretty much the whole animal — the same high protein, high calorie meals that Lewis and Clark ate when game was plentiful.

By comparison, even when an agriculturalist's crops provided enough calories to meet their energy requirements — which probably were as high if not higher than those of hunter-gatherers because farmers may work every day from dawn till dusk — they didn't provide as much protein or calcium as the hunter-gatherers' diets did. Even nuts generally have more protein per ounce than grains, especially domesticated grains. That's why agriculturists also looked to legumes — beans, peas, and even peanuts — for their protein, much as modern vegetarians do. But it's a challenge to get enough protein from nonanimal sources, and it was even harder in the era before supermarkets.

Nutrition continues to correlate with human height today. The economist Richard Steckel compared the height of people living in Europe from the ninth through the nineteenth century by measuring the length of their thigh bones, or femurs, which account for about 25 percent of our height. He estimated that an average man in the early Middle Ages stood 5 feet 8¼ inches tall. But by the seventeenth and eighteenth centuries, when more people lived in cities, where there was less vitamin D and fresh meat, the average man's height had declined to 5 feet 5¾ inches — a reduction of 2½ inches. To put this into perspective, it's about the same as the current difference in height between an average young man in Canada and an average young man in urban China. And by the early twenty-first century, better nutrition explains why the average height of young men in affluent European countries increased to about 5 feet 10 inches.

In most societies, women are generally about four to five inches shorter than men, regardless of the height of the men. In this context, the 8½-inch height difference between Paleolithic men and Paleolithic women was noticeably larger than the sex difference we've seen ever since. Perhaps the higher absolute and relative difference in height between hunter-gatherer men and women is because the women didn't fully share in the protein bounty. Another possibility is that women became pregnant at a young age, when they were still growing, and their growth would have been stunted if they had had to share their protein and calcium with a developing fetus.

But just as the average heights of northern Europeans varied over the centuries along with their diets, modern population-wide heights can also change rapidly. In China, for example, relatively better-fed urbanites were 1½ inches taller than their less-well-nourished rural counterparts in 2002. And although 7-foot-6-inch Yao Ming, the internationally acclaimed basketball player, is surely an anomaly, the increasing heights of better-fed young urban Chinese shows how quickly improved nutrition can translate into taller stature. Perhaps the tallest people today are the Dutch, who are, on average, the best nourished and healthiest of the people living in Westernized nations, and the Masai of West Africa, who subsist on a high-protein diet based on the blood, milk, and occasionally meat of their cattle.

Obesity Through History

For our ancestors, obesity couldn't have been common and certainly would have been detrimental. Whether you're hunting, foraging, planting, or herding, you have to work harder if you weigh more. And the extra weight slows you down — a problem whether you're chasing after your dinner or it's chasing after you!

Nevertheless, we know that our Stone Age ancestors accurately depicted massively obese women in ivory in Germany around 35,000 years ago and in limestone in Austria about 25,000 years ago. But we don't know whether these were fertility symbols, curiosity pieces, or idealized models of womanhood. By 8,000 years ago, many figurines depicted people of reasonable body proportions, but some statuettes, especially the famed "Mother Goddess" figurines found in Turkey, were still markedly obese.

By the time of the Egyptians, we see well-proportioned drawings of slim young people, but we also see progressively heavier and even potbellied older royals, including the stone relief of a plump man being fed by a thin servant in Mereruka's tomb. Some Egyptian mummies have characteristics consistent with obesity. And affluent Egyptians, according to Herodotus, periodically fasted or even induced vomiting to avoid gaining too much weight. Artifacts of the Maya, Inca, Aztec, and Mesopotamians also include obese human figures.

About 2,500 years ago in Greece, Hippocrates warned against the adverse health consequences of excess weight. The sweet taste of diabetic urine and the propensity for fat people to be afflicted by diabetes was described by Hindu physicians 2,500 years ago. And we know that being overweight or even obese was often a marker of affluence in ancient Rome, even though Galen, the most famous of all Roman physicians and the personal doctor to the Roman emperor Marcus Aurelius about 1,800 years ago, advocated exercise as well as dietary discretion. Upper-class Roman obesity was typically ascribed to overindulgence and laziness, in many ways symbolizing the internal rotting that led to the empire's decline.

Obesity is also driven by perceptions of social norms. Rubenesque women were considered attractive in the affluent seventeenth-century world of upper-class Holland, but probably because their relative corpulence so contrasted with the body shape of peasants.

Perhaps the most striking and relatively recent example of societal influences is the weight of American presidents. From 1884 to 1912, six consecutive American presidents—Grover

Cleveland, Benjamin Harrison, Grover Cleveland again, William McKinley, Theodore Roosevelt, and William Howard Taft— were all obese, topped off by the 340-pound Taft. During this era, affluent American city dwellers had unprecedented access to food grown by others as well as to mechanical transportation. Since these affluent individuals lived longer than contemporary poor people, who were typically undernourished, the adverse effects of obesity itself were initially underappreciated.

Except for Zachary Taylor, no other American president, before 1884 or after 1912, has ever been obese. Starting with Woodrow Wilson, our presidents and our well-educated elite began to recognize at least some of the side effects of obesity. As Wallis Simpson—whose engagement to King Edward VIII precipitated his abdication of the throne—famously said in the 1930s, you "can never be too rich or too thin." Today, the Americans least likely to be obese are medical doctors.

This changing social norm for weight is perhaps no different from the social norm for tanning. The women of the pre–Civil War US plantation aristocracy were prized for their pale skin, which proved they didn't need to work in the sun—a phenomenon echoed in modern Japanese and Chinese fashion trends. By comparison, twenty-first-century Americans pay to go to tanning salons because tanned skin symbolizes youth and leisure time spent outdoors, especially at the beach.

But let's reflect for a moment. Our presidents were fat when our population was thin, and our presidents are normal weight now that our population is fat. Around the turn of the twentieth century, obesity was common only among the most privileged of Americans—such as our presidents—just as it's now become an epidemic in the newly affluent classes of developing countries. By comparison, in twenty-first-century America, middle class and poor people—and especially women and children—are, on average, heavier than affluent people.

Why, you might ask, didn't the average American follow the

bad example of these overweight presidents and become obese in the early twentieth century? After all, presidents weren't the only ones who expended fewer calories with the advent of industrialization, urbanization, and mechanized transportation. The likely reason, as explained by global nutrition expert Boyd Swinburn and his colleagues, is that caloric consumption, especially from wheat, declined in America from about 1910 through about 1960, perhaps because of a rapidly growing population, two wars, and the Great Depression. But by the 1970s, the ready availability of inexpensive high-sugar, high-fat foods led to progressive increases in caloric intake. Food is now so plentiful that the amount we throw away has gone up by about 50 percent in the past 30 years or so, even though we're eating more. Not only was Thomas Malthus wrong in 1798 when he famously predicted that the world wouldn't produce enough food to feed a growing population, the approximately 1,200–1,300 calories' worth of food that we now throw away each day per person in the United States would also be enough to feed millions more people—or, as our parents used to say when we didn't finish our dinners, all those starving children in Asia and Africa.

Why Do We Gain Weight?

For most of human history, potential starvation was never more than several weeks away, and obesity wasn't even on the radar screen. Eating patterns were naturally linked to physical activity levels, and human traits—hunger, taste, digestion, absorption, metabolism—were geared to protect us from either gaining or losing much weight. But the thermostat was always set to protect against the biggest threat to the survival of the species—starvation. And, like the Pima Indians in the arid Southwest, our ancestors repeatedly faced those threats. The better they were at meeting that challenge, the more likely they were to pass on a variety of traits that ensured we never ate too little, could always eat too much, and could store fat immediately to protect ourselves from starvation later.

We gain weight for a simple reason: we eat more calories than we burn. The best way to assess whether you're overweight is to calculate your body mass index (BMI). The formula is simple:

$$\frac{weight\ in\ pounds}{(height\ in\ inches)^2} \times 703 \quad or \quad \frac{weight\ in\ kilograms}{(height\ in\ meters)^2}$$

A normal BMI is between 18.5 and 24.9. You're classified as overweight if your BMI is between 25 and 29.9, obese if it's 30 or above, and morbidly obese if it's 35 or higher. So for me, just a bit over 5 feet 9¼ inches tall, a normal weight would be between 125 and 168 pounds, overweight would be 169–202 pounds, and obese would be 203 pounds or more. This formula can't distinguish a potbelly from six-pack abs, and some athletic people may be slightly overweight because of extra muscle. But it's close enough, on average, to define most of us individually and all of us collectively.

For those of us who have a low basal metabolic rate or don't get a lot of exercise, it's easy to overeat. But our caloric needs actually increase as we gain weight, in part because it takes more calories to maintain a bigger person's temperature and organs at rest, and in part because it also takes more calories to move a bigger person around. And, as noted by James Hill, an obesity expert at the University of Colorado, and his colleagues, the result is that heavier people don't gain nearly as much weight as you would think based on their increased food consumption. Here is a striking example. In the United States, the increase in average adult daily caloric intake over the past 40 years would theoretically translate into an average weight gain that would be 30 to 80 times as much as the actual 20 or so pound average gain! Thankfully, our heavier bodies burn most of our extra calories—so while we've become overweight or even obese as a population, we haven't become blimps.

But let's go from the population to the individual. One pound of fat stores about 3,500 calories. That theoretically means that if you ate just ten more calories than you need per day—less than 1 percent more than your body actually needs—you would gain about a pound per year. However, analyses by the nutritional physiologist Kevin Hall and his colleagues estimate that you would need to eat an extra ten calories per day for three years to gain and maintain an extra pound of weight: about half that pound would be gained in one year and the other half would be gained over the subsequent two years. That means you'd have to consume about 11,000 extra calories overall, including what it would take to put on the weight and what it would take to carry it around, to have a sustained one-pound weight gain. And to gain about a pound every year throughout adulthood, you'd have to keep raising your intake (or reducing your expenditure of calories) every year by an incremental ten calories over the prior year. These estimates show that we can slowly gain weight by always eating ten more calories per day than we did in the prior year, year after year, or by always burning just ten fewer calories each day than we did the year before.

But even if you go from ten extra calories per day to 50 or so extra calories a day—enough to lead to obvious weight gain—it may be hard to recognize that you're really overeating. That's because those 50 calories could be consumed by eating about one Oreo cookie, two Hershey's kisses, a small apple, a mandarin orange, seven almonds, half a banana, one sausage link, half a slice of bread, or a tablespoon of honey. It might not seem like much each day, but the calories would add up, and that's why many Americans gain close to a pound a year.

Of course, it's important to remember that individuals vary dramatically. In one study of 12 pairs of identical twins from diverse backgrounds, feeding them 1,000 calories more than they needed always resulted in weight gain. Although any one twin always gained about the same amount of weight as the other, some pairs of twins gained three times as much weight as other pairs of twins. These findings demonstrate the marked variability in how individuals gain weight as well as the fact that much of the variability is genetic.

Another practical challenge is that it's hard to tell if you've gained a real pound of fat. Weight can easily vary by two or three pounds based on how much salt we've eaten and, as a result, how much water we've temporarily retained. That's why we hate to weigh ourselves after a salty dinner. So it's difficult, even with the best of scales, to monitor our weight with the precision needed to be sure we aren't gradually gaining weight until we've put on several pounds.

As we think about how easy it is to gain weight gradually over the years, let's consider a different question. Why isn't it just as easy to go in the other direction and gradually lose a pound every three years or so? Of course, we rarely see an otherwise healthy person slowly lose weight voluntarily year after year, even though you might think it would be just as easy to undereat by ten calories per day, year after year, as it is to overeat by ten calories per day.

There are several reasons why it's actually harder for a healthy person to lose weight than to gain weight. First, as we lose weight, our caloric needs decline—we need about 20 fewer calories for each 1 percent of weight that we lose. In addition, when we lose weight, regardless of how much we weigh, at least seven different hormones and molecules that stimulate our appetite get revved up and remain elevated for years—a key survival trait for our ancestors but a major obstacle for those of us trying to lose weight. If we need fewer calories and we're hungrier than we were before, it's very hard to lose weight, no matter how much you weigh. These internal forces that defend our weight also explain why it's actually easier to avoid gaining weight in the first place than it is to lose it once you've gained it.

So let's review for a minute. Our bodies aggressively protect us from losing weight (by reducing energy expenditures and slow-ing metabolism) and try to keep us from gaining weight (by increasing our energy expenditure). This intelligent design tries to keep us at a reasonable weight, though the downside protection is much, much more effective than the upside protection.

Since a heavy person burns more calories than a light person, you might think that this self-correcting mechanism would mean that we would each reach a ceiling where our overeating compared with recommended caloric intake would simply balance the increased energy burned by a fatter body. But many people continue to gain weight, in part because they may progressively eat more but also because they progressively exercise less. Because of their decreased physical activity levels, weight gainers need fewer calories than they would if they maintained their prior exercise levels. Work done in the 1950s by the famous dietary expert Jean Mayer showed that humans are really good at matching caloric intake to energy expenditures above a certain threshold of physical activity. But if our physical activity falls below that threshold, we're not well balanced at all — we tend to overeat relative to what we need. This intrinsic trait undoubtedly protected the weight of prehistoric humans who were temporarily incapacitated and unable to go out to hunt and gather — so they could maintain enough strength to resume these activities after they recovered.

With many of us apparently below this physical activity threshold, it's easy to eat more than we need, and that means we really must either increase our activity levels or restrict our food intake — or both — if we want to avoid gaining weight. James Hill and his colleagues call declining physical activity levels, which began with the Industrial Revolution and continue today, the "permissive" factor in the modern obesity epidemic — or, to use another metaphor, the reduction of physical activity loaded the gun, and the recent overavailability of cheap calories pulled the trigger.

This permissive concept is certainly consistent with what happens in the rest of the animal kingdom. Animals in the wild don't get fat, as they stalk their prey or graze on the open range, whereas animals living on farms or in zoos can easily be fattened up when ranchers and zookeepers reduce their activity levels and provide an oversupply of enriched foods. That's why the meat we eat today is so much fattier—and loaded with the saturated fats that raise our cholesterol levels—than the meat eaten by hunter-gatherers.

When we become overweight or obese, where do we store all the excess? The answer is simple: in the fat cells throughout our bodies, but especially in our abdomens, waists, and hips. Scientists estimate that those of us who are thin may have about 35 billion fat cells, which collectively store about 130,000 calories—enough for us to survive for about the same length of time as those Irish hunger strikers. But those of us who are obese can have as many as 140 billion fat cells, each of which may be twice as big as the fat cells in a thin person. These bigger and more numerous fat cells can store up to one million calories—enough for perhaps 16 months of starvation, even more than the aforementioned 450-pound man's world record for going without calories.

From an evolutionary perspective, the ability to store fat was undoubtedly critical to the survival of our species, just as it's critical for so many other mammals. In the wild, many animals fatten up in anticipation of winter or hibernation, an annual gorging that provides them with an undeniable survival advantage. Since our ancestors, at least in cold climates, would also have faced food shortages in winter (especially before they used salt to preserve food), they, too, may have fattened up as the temperature began to fall. And we would expect that those who were the best at getting enough calories and storing some of them as fat were the ones most likely to survive the winter as well as any subsequent food shortages. Now, however, with refrigeration, overstocked supermarkets, pizza, and Happy Meals, most of us never face such shortages—but we still have genetic traits that make us act as if we do.

Why We Should Care About Obesity

If obesity were nothing more than a cosmetic problem, all the attention given to it would be dismissed as much ado about nothing. But that's not the case.

In the United States, obesity reduces life expectancy by about six years — somewhat less if your BMI is just above 30 and up to ten years as it hits 40. Compared with people of normal weight, obese people are at much higher risk for diabetes, high cholesterol levels, heart disease, high blood pressure, and even cancer, presumably because their higher insulin levels serve as a growth stimulant for all cells, including cancer cells. In the United States, economists estimate that an obese person costs at least an extra $1,400 per person per year in medical care (or nearly $150 billion in aggregate) and a diabetic person costs an extra $8,000 (about $250 billion in aggregate).

Although there's been some controversy about whether a little extra weight really is as bad as we used to think, the best analyses are careful to look at all your risks so that the actual independent effect of weight itself can be determined. In these analyses, the risks of death increase as weight rises.

Approximate Risk of Death Versus Risk at Normal BMI of 20–24.9

BMI		Risk
25–29.9	Overweight	10–15% increase
30–34.9	Obese	45% increase
35–39.9	Very obese	90% increase
40–49.9	Morbidly obese	150% increase

Such careful analysis is critical, since some people are thinner because they're less healthy (they smoke or they're otherwise ill) and some people weigh more for reasons that may bring secondary benefits (they're more affluent or they get better medical care). For example, being slightly overweight seems okay for people with cancer and heart disease; that's probably because overweight cancer patients aren't yet as sick as their lower-weight counterparts and because overweight heart patients may be taking more aggressive medications for their cholesterol levels and blood pressures. That's also why, at first glance, obesity didn't seem so bad for turn-of-the-twentieth-century American presidents, and it doesn't seem so terrible in developing countries, where it's associated with the other benefits of disproportionate affluence. But the affluent obese in developing countries have higher mortality rates than affluent men and women of normal weight, just as the nonaffluent obese in developed countries have higher mortality rates than their normal-weight counterparts.

Ultimately, the takeaway is to keep your weight in the normal range, preferably in the higher half of normal (a BMI of about 22 to 24.9) — especially if you are at risk for or already have a problem such as cancer, which could cause poor appetite and weight loss if you get sicker.

Why We Should Care About Diabetes

As a direct result of the obesity epidemic, an estimated 20 million Americans — about 10 percent of us, including more than 25 percent of Americans over age 65 — have diabetes. The worldwide average is lower — about 3 percent overall and about 12 percent for people over age 65 — but rising, and many parts of the world, especially parts of Polynesia, the Middle East, and the Caribbean, are already ahead of us.

If obesity is bad, the diabetes it can cause is even worse. Diabetes is more than just a chemical abnormality. It reduces life expectancy by about 12 years, and this risk is independent of the risks associated with obesity itself. Diabetic people have more heart attacks, strokes, kidney failure, blindness, leg amputations, and infections of all sorts than do nondiabetics, independent of their body weight.

There are two major types of diabetes, unimaginatively termed *type 1* and *type 2*. Type 1 is sometimes called *juvenile diabetes,* because it usually appears in childhood and adolescence, or *insulin-dependent diabetes,* because it results from a complete or almost complete absence of insulin production as a result of immune damage to the pancreas, which makes insulin. Type 2 diabetes is also called *adult-onset diabetes* because it commonly appears in overweight, middle-aged, and elderly individuals, or *non-insulin-dependent diabetes* because it's associated with normal or, more commonly, elevated levels of insulin. The problem in type 2 diabetes is not a shortage of insulin but rather a relative insensitivity to its effects.

When we talk about the diabetes epidemic throughout the world, we're really talking about type 2 diabetes. And based on current dietary trends, it's estimated that an American born in 2000 has about a 35 percent probability of getting diabetes, mostly type 2 diabetes. Even worse, however, is that it's no longer just adult-onset. Now we see it increasingly in obese young adults, adolescents, and even children. And as type 2 diabetes gets more severe, it may no longer be insulin-independent. Patients often stop responding to medications that are designed to make them more sensitive to insulin because their bodies become progressively less and less sensitive to the insulin itself. Or they may be unable to increase their own insulin production, even when given stimulating medications, because their pancreases "burn out" and can no longer respond. Instead, they may actually need to take insulin, just as people who have insulin-dependent type 1 diabetes do.

Effective medications can reduce the blood sugar level and even get it down to normal, but very aggressive attempts to control blood sugar in obese diabetics may actually increase, not decrease, complications because of the dangers of *hypoglycemia,* a very low blood-sugar level. Acute hypoglycemia can cause confusion or even coma because of the brain's reliance on glucose for its energy needs, and recurrent episodes can cause residual brain problems. That's why diet and exercise, though difficult to achieve, remain cornerstones of effective therapy.

Not everyone who gains weight or even becomes obese develops type 2 diabetes, and some type 2 diabetics aren't overweight. A number of genes seem to influence who gets type 2 diabetes: typically, these are genes that affect how our bodies handle sugar and respond to insulin, and they're usually but not always related to weight gain.

Obesity, diabetes, and their associated diseases now kill far more people worldwide each year than starvation or undernutrition do. The Pimas of the twentieth century aren't just a curiosity; instead they turned out to be an early peek into the future of humankind. Our genes, which were so well adapted to help our ancestors survive in varying environments, each with its own caloric challenges, are ill suited to a sedentary world with a plentiful food supply.

Learning from the Pimas

To understand how the Pima Indians became the poster children for the epidemic of obesity and diabetes, it's helpful to understand their history. The Pimas are thought to be direct descendants of the first group of Asians who migrated across a Bering Strait land bridge to North America about 30,000 years ago. They settled near the juncture of the Gila and Salt Rivers in the Sonoran Desert of Arizona, where they depended on a sophisticated irrigation system and watertight woven baskets to transport their limited water supply. Their traditional diet, based on agriculture and river fishing, persisted until their water supply was diverted to support upstream farmers in the late nineteenth century. To avert looming malnutrition, the US government brought in flour, sugar, and lard. The Pimas began frying the flour dough in lard, creating a new staple food and raising the fat content of their diet from about 15 percent to about 40 percent.

Based on the early findings of Ales Hrdlicka and Elliott Joslin, there was no particular interest in the Pimas' diet or concern about their health until the 1950s, when they were found to have remarkably high rates of gallstones and cancer of the gallbladder. In an attempt to understand this strange predilection, which even today is not well explained, scientists began to study their diet in detail.

As part of these studies, scientists noted that the Pima diet in the early 1950s averaged about 2,800 calories per day—approximately the recommended daily intake for active Americans at that time but already far too many calories for people who no longer hunted, gathered, farmed, or had a tradition of vigorous leisure exercise. As a result, about 12 percent of the newly sedentary Pimas had diabetes—around ten times their rate in 1937. Unfortunately, the Pimas' daily caloric intake increased to about 3,200 calories per day at the same time that their recommended daily intake, based on their physical activity, declined to closer to 2,000 calories per day. Of course, the Pimas gained a lot of weight—on average, they were about 50 percent heavier than their desirable weight by 1970. And their rates of diabetes went up even more.

We can also compare the Pimas of Arizona to their close relatives, the Pimas of Mexico. In the mid-1990s, the Arizona Pimas' daily intake of about 3,000 calories was 500 calories per day higher than that of their Mexican counterparts, even though their physical activity was 80 percent lower. Not surprisingly, the Arizona Pimas outweighed the Mexico Pimas by an average of about 65 pounds—fully 40 percent! And as a result, the rate of diabetes among the Arizona Pimas was six times higher.

Although it is easy to understand why the Arizona Pimas,

based on their low level of physical activity and high caloric intake, have a higher rate of obesity and diabetes than their Mexican cousins, let's ask a different question. Do they have more obesity and diabetes than other people with similar exercise and dietary habits?

To address this question, let's consider two hypotheses. The first is the *thrifty genotype hypothesis,* which proposes that some of us can get by on fewer calories than other people need either because we need fewer calories to power our resting organs (lower basal metabolic rates) or because our muscles are more efficient and need fewer calories to power our bodies. The second is the *thrifty phenotype hypothesis,* which suggests that undernourished babies become conditioned to be more metabolically efficient from their time in utero and will be especially at risk for obesity and diabetes if later exposed to a plentiful Western diet.

The thrifty genotype hypothesis, which was first proposed by geneticist James Neel in 1962, argues that the genetic advantage for surviving food shortages causes obesity and diabetes in times of plentiful food. Simply put, if all of us were built to err on the side of eating a little bit more than we need whenever we can get it, it means that some of us undoubtedly come from ancestors who faced particularly challenging food shortages — people for whom this trait was therefore even more critical. And the twin studies we discussed earlier show that 55–85 percent of the variations in human weight are genetic.

If multiple mutations explain the appearance of adult lactose tolerance beginning about 7,000 years ago and light skin pigmentation beginning perhaps 60,000 years ago, imagine how many different mutations might explain a heightened ability to maintain weight and store fat over the 2.3 million years of the *Homo* genus. Not surprisingly, we already know at least 30 different genes that are associated with weight. One well-studied example is a mutation in the *FTO* gene, which makes a protein that is distantly related to adrenaline; the mutation slows down the rate at which we burn fat. If you get the mutation from one parent, you'll be about three to six pounds heavier, on average, and if you get it from both parents, you can be as much as ten pounds heavier. It increases the risk of becoming obese by about two-thirds, and, interestingly, it also reduces the risk of depression by 8 percent. About 15 percent of Pimas have a mutation in this gene, compared with about 45 percent of Caucasians.

Variations in the *SIM1* gene, which is critical for the development of the part of the brain that controls the appetite, have also been associated with obesity in the Pimas and other populations. And mutations in *MRAP2,* which also affects how the hypothalamus functions, can affect appetite as well as how much fat we store independent of our caloric intake.

In addition to increasing the risk of weight gain and obesity, some genes seem to cause type 2 diabetes regardless of weight gain, presumably because they affect our metabolism. *HNF1A* gene mutations, which increase the risk of type 2 diabetes about fourfold, are found in about 2 percent of Mexicans with diabetes and are not found in non-Latino populations. Even more impressive is a gene called *SLC16A11,* in which mutations — which occur in 50 percent of indigenous Mexicans and Latin Americans, about 10 percent of East Asians, but rarely in Europeans or Africans — increase the risk of type 2 diabetes by about 25 percent if you inherit it from one parent and by about 50 percent if you inherit it from both. This gene influences how efficiently our fat cells store a concentrated fat called triacylglycerol — the same high-energy fat that hibernating mammals use to get through the winter and that birds use as their source of energy when they fly long distances without eating. But it's especially interesting because this gene originally was one of those Neanderthal genes that, as I described in chapter 1, made its way into our DNA after our ancestors left Africa. It seems that the same gene that helped the Neanderthals store concentrated fat so they could survive cold winters also helped our ancestors survive as they moved east across Asia, especially when they crossed the Bering Strait into North America.

We also know that the various genes currently known to increase the risk of obesity-related diabetes are more common in Africa and generally less common the farther away you are from Africa. So it's not surprising to see that the rising leisure class in sub-Saharan Africa is mimicking the Pima experience and the obesity of our US presidents from 1884 to 1912. On a recent visit to the health ministry of one of the poorest countries in the world, with near-record levels of HIV infection, I met with people responsible for promoting the health of the country. Four of the ten people who met with me were obviously obese, and the single thin person had purposely lost weight after being diagnosed with obesity-related diabetes some years earlier.

If a thrifty genotype provides a survival advantage, the obvious question is why don't we all have it? One postulate is that lactose-tolerant people more easily got their required calories from dairy products, and this mutation protected them even if their other genes weren't so thrifty. This hypothesis fits with the absence of the Neanderthal-derived *SLC16A11* gene in modern Europeans and the somewhat lower rates of diabetes in milk-drinking people of northern European origin.

The thrifty phenotype hypothesis, which I'll discuss more in chapter 6, was first proposed in the 1990s, when researchers C. Nicholas Hales and David Barker noted that low-birth-weight English babies were more likely to develop diabetes later in life than were normal-weight babies. A similar phenomenon was reported in Nauru, an island in the Pacific, which now has the world's highest rate of diabetes, and among undernourished Ethiopian Jews who migrated to Israel.

How about future generations? Among Nauruans, the risks of obesity and diabetes have declined somewhat in recent generations, now that infants are better nourished, but Nauruans are still a lot heavier than their own ancestors or people in most other parts of the world.

For reasons we don't fully understand, parental overnutrition also increases the risk of obesity in their children. Mothers who gained a lot of weight during pregnancy also tend to have obese children — perhaps because they pass on thrifty genes, perhaps because of postbirth environmental factors, but also perhaps because in utero exposure to high insulin levels permanently influences babies' lifelong metabolism.

The scores of genes already implicated as contributors to obesity and diabetes, as well as others that are likely to be discovered in the future, confirm that our ancestors developed many thrifty genotypes to protect against starvation. However, none — by themselves or in combination — can yet provide a genetic "smoking gun" to explain exactly which lactose-intolerant Pimas don't become obese diabetics and which lactose-tolerant people of European descent do.

The Modern Conundrum

To survive despite an always unreliable food supply, the human body evolved to outwit the recurrent threat of starvation. Our taste buds stimulate our desire for calorie-dense fat, sugar, and protein. And our intestinal tracts maximize the extraction of nutrients, especially from foods that must be broken down from their natural forms. In addition, we're programmed to overeat whenever possible so we can store fat to protect against future shortages. However, our ancestors didn't want to get too fat, since heavier people actually need to consume more calories to do the same work as their lighter and fitter counterparts. Although our ancestors never ate a uniform Paleolithic diet, we do know what they didn't eat. Remarkably, about 70 percent of our modern industrialized diet comes from foods that weren't available to hunter-gatherers:

Modern Western Diets: Calorie Sources Not Readily Available to
Hunter-Gatherers in the Pre-agriculture Era

Source	Approximate Percentage of Calories
Animal-derived dairy products	10
Cereal grains	24
Refined sugars	18
Refined vegetable oils	17
Alcohol	1
Total	70%

We can't dismiss the importance of genetic predispositions toward obesity, but the modern obesity epidemic certainly isn't caused by a sudden change in the worldwide human genome. Simply put, the problem is that we're consuming more calories and exercising less. The effects of these behavioral changes in just the last two generations are exemplified by the *FTO* gene we just discussed. In one large study, it didn't predict body weight in people born before 1942, but it was a significant predictor for people born later than then. These results demonstrate how genes that have been protective for 10,000 generations can become detrimental to our health in just two generations.

In 1960, only about 20 percent of Americans were overweight, and obesity was uncommon. Starting in the 1970s, however, the intake of refined carbohydrates and fats, and hence the consumption of calories, increased dramatically.

By how much did average caloric intake increase? It's hard to know for sure, because we probably underestimate how many calories we eat per day. National food surveys suggest that the average American adult claimed to eat about 2,100 calories per day in the 1970s and almost 450 more calories per day in 2010. By comparison, national data on food production and average weight suggest that, even after accounting for food waste, adult caloric consumption probably did increase by about 500 calories per day—but from about 2,400 calories per day in the 1970s to about 2,900 per day now. And this increase comes from eating more fats and carbs, in bigger portions and more often.

Now nearly 35 percent of American children and adolescents and two-thirds of American adults are overweight, and about half of them—17 percent of children and adolescents and more than one-third of adults—are obese. Worldwide, about 500 million adults, more than 10 percent of us, are now obese—double the rate of 30 years ago—and another 1.5 billion are overweight. Obesity rates range from less than 5 percent in poor countries such as Ethiopia to 20–35 percent in much of the Western developed world (with the rate in the United States higher than any other high-income country) to more than 60 percent in the Pacific island countries of Samoa, Tonga, and Nauru. More sub-Saharan mothers are now overweight (19 percent) than underweight (17 percent), and 5 percent of them are obese. We see the same trends in tropical, temperate, and Arctic climates, with the largest increases in people who made the biggest transition from their traditional diets to Western diets, regardless of whether their traditional diets were plant-based or, like that of the Inuit, almost entirely animal-based. Why? Because none of them previously included an oversupply of sweets and high-density carbohydrates.

The only affluent country with a persistently low rate of obesity is Japan, at less than 5 percent. In Japanese culture, especially in Okinawa, the Confucian tradition of hara hachi bu instructs people to stop eating when they are 80 percent full. Such a societal norm is a great way to be sure a limited food supply is shared so it can benefit more people. In 2008, Japan also instituted a national law requiring all adults between the ages of 40 and 70 to have their waistlines measured each year: the law says the measurements should be no more than 33.5 inches for men and 35.4

inches for women. People with larger waistlines and any weight-related problems must lose weight or receive mandatory dietary counseling.

These observations further support the hypothesis that a decline in physical activity sets the stage for obesity and that the ready availability of cheap calories makes it happen. The same thing happened to the Pimas, who could no longer be active farmers when their water supply was constrained and whose staple food suddenly switched from what they grew or gathered to the calorie-dense flour and lard that they were given.

But even as we face an obesity epidemic in the Western—and Westernizing—world and as we see pockets of overweight and obesity breaking out in the developing world, the historic survival benefits of our eating habits and metabolic traits shouldn't be underappreciated. Malnutrition still contributes to the deaths of nearly three million children each year—almost half the world's childhood mortality. About 70 percent of the world's 150 million underweight children live in just ten countries, half of them in South Asia. But things are changing very rapidly throughout the world. Childhood undernutrition was the leading worldwide cause of disease and disability in 1990, but it was surpassed by both overweight and diabetes by 2010.

So food isn't in short supply in the modern industrialized world—it's overabundant. And it's overabundant in tasty concoctions that are loaded with fats and sugars that humans find hard to resist. To make matters worse, our feeling of satiety is substantially dependent on the volume of food we eat. We'll now eat more calories from densely nutritious modern food before we feel full than if we tried to eat the same calories from the usually less calorie-dense food of the Paleolithic diet.

Obesity certainly isn't new, but what is new is its high and rising frequency. And its sociology is bimodal—marked increases in the developing world as a side effect of at least relative affluence, and equally marked increases in the less-affluent developed world, where people are much less active than they used to be and, simultaneously, have ready access to massive amounts of cheap calories. The US obesity epidemic—or, for that matter, the worldwide obesity epidemic—is unlikely to disappear unless we suddenly eat a lot less and exercise a lot more. We're simultaneously fighting a number of social and behavioral challenges. First, in much of the world there's still a strong social association between being overweight, even obese, and being affluent and successful, going back at least 35,000 years and at least 1,750 generations. Second, even though most fashion magazines equate thinness with beauty, American middle-class and underclass social norms have changed to accept obesity and even to regale "full-figure" beauty. And third, our exercise levels have declined, often to below a threshold that allows us implicitly to balance our eating with our activity level.

But the biggest problem is that we're fighting our genes. All the remarkable traits that helped our ancestors survive recurrent food shortages are simply out of sync with modern supermarkets, fast-food restaurants, and unlimited snacking options. The Pima Indians really never were the outliers they were first thought to be. Unfortunately, their obesity and diabetes were a profound early warning of a new global epidemic in a world that's changed far faster than our genes can.

CHAPTER 3

Water, Salt, and the Modern Epidemic of High Blood Pressure

On April 12, 1945, vice president Harry Truman had just arrived in the office of Sam Rayburn, the Speaker of the House, for their usual late-day bourbon when he was told to call the White House. The caller asked Truman to come promptly to the president's private residence, where Eleanor Roosevelt met him and said, "Harry, the president is dead." After a brief pause, Truman reportedly asked Mrs. Roosevelt, "Is there anything I can do for you?" As the story goes, Mrs. Roosevelt replied, "Is there anything we can do for *you? You're* the one in trouble now."

It's well known that Franklin Delano Roosevelt contracted polio in 1921 at the age of 39. But it's not widely known that, throughout his presidency, Roosevelt suffered from what was then an inexorably progressive, untreatable chronic disease — high blood pressure. His death at age 63, nearly 25 years after the polio was diagnosed, wasn't related to his polio. Rather, he died from a sudden stroke when his severe high blood pressure literally blew a hole in a small artery in his brain, causing massive bleeding inside his head and killing him within hours.

Although the causes of high blood pressure weren't known in 1945, we now recognize that they're the same biological mechanisms that protected our ancestors from debilitating or even deadly dehydration. In the Paleolithic era, humans needed more calories than we do now to sustain their higher levels of exercise. In addition, they also needed more water and salt because they sweated more, especially with their active lifestyle in our original African habitat. Since their sources of water and salt were far less predictable than they are now, their bodies had to be built to seek, consume, and retain enough of both to help protect against future shortages. And when water and salt were limited, they needed a variety of hormones to keep dehydration from dangerously lowering their blood pressures. Put simply, Roosevelt died from the unintended consequences of an overprotective set of behaviors and hormones that had been critical for the perpetuation of the human species for 200,000 years.

THE SURVIVAL CHALLENGE

By weight, our bodies are about 60 percent water. Just as we must eat food in order to supply the calories that our bodies need for energy, so, too, must we drink water, an essential component of our blood as well as of all the cells in our body. Unlike food, which is burned to provide energy, water is lost through one of four routes. At least a quart or so is lost each day in our urine, where water is needed to help flush out the waste products of our daily energy metabolism — waste products that otherwise would quickly build up in our bloodstreams and kill us. We lose water from our skin, in part because it can never be watertight and in part because we sweat. We also lose water just by breathing, because our lungs have to humidify the air we inhale to keep it from being too irritating. And finally, we lose a little in our feces. Even without any detectable sweating or diarrhea, we lose an average of a little more than half a quart of water from our skin, exhaled breath, and feces every day—so combined with the water in our urine, we normally lose a total of at least a quart and a half or so per day.

The additional water we'll lose by sweating depends on the heat our bodies produce and on the external temperature. The more we exercise, the more calories we burn as energy, the more heat we produce, and the more we need to sweat to avoid becoming overheated, especially if it's hot wherever we're exercising. Overall, adults in modern industrialized countries usually sweat enough so that their total water loss will be at least 2–2.5 quarts of water per day. Since food provides about half a quart of water per day, we have to drink at least 1.5–2 quarts per day to meet our minimum daily losses. And because we should have a margin of safety in case we sweat more than usual, the generally recommended daily fluid intake is 3.5–4 quarts per day for men and 2.5–3 quarts per day for women.

If we don't drink enough water, we get dehydrated—and our blood pressure falls. Why? It's really no different from what generates the pressure at the end of a hose. When we get dehydrated, our blood volume falls because we have less liquid plasma, in which our red blood cells are transported. At first, our arteries constrict to maintain blood pressure despite lower blood volumes and blood flow, just as a hose would if we used a nozzle to maintain water pressure despite less water coming from the spigot. But at some point, pressure falls. And when blood pressure falls, we'll get light-headed, faint, or even die because not enough blood gets pumped to our brains and ultimately to other vital organs.

Without food, even healthy people will starve to death, usually within about seven to ten weeks, as we cannibalize our stored fat and even muscle to supply the calories we need. By comparison, we'll die within several days without water, because we don't have a large internal tank in which to store weeks' or months' worth of excess water.

But pure water alone can't keep us alive. Since we also lose varying amounts of salt in our urine and our sweat, we need to eat salt. And just as our kidneys help us get rid of any excess water, they also help us get rid of just the right amount of salt so our bodies don't get either too diluted or too salty. Salt also helps us hold on to needed or even a little extra water. This phenomenon is obvious to people who weigh themselves daily—an especially salty meal makes us gain weight for a day or two because we temporarily retain more water to balance the salt.

BACK TO THE PALEOLITHIC ERA

Our historic need for water is especially linked to a key human survival trait—endurance. Endurance, or the ability to keep moving when other animals fatigue, has been critical to the perpetuation of the human species. Most of the wild animals that humans have historically eaten are either bigger or faster (or both) than we are, so how could they be caught? Humans survived in part by capturing the weak and the young—the same strategy used by many other carnivorous predators. But these other carnivores are usually sprinters that run faster or are far stronger than their prey. We, on the other hand, must either outsmart our prey or outlast them. Prehistoric hunting often required groups of our strategic, large-brained ancestors to run an animal into exhaustion, corner it, and kill it, thereby neutralizing whatever speed or size advantage the animal may have had.

As Christopher McDougall notes in his book *Born to Run,* our legs and feet are beautifully designed for running long distances. Daniel Lieberman, an evolutionary biologist, emphasizes that we run best when barefoot, landing on our toes rather than our heels, and that the endurance needed to run down prey was key to our survival. And in his book *The Art of Tracking,* Louis Liebenberg describes how the Kalahari Bushmen of Botswana hunt antelope. First they hydrate with water. Then they take turns running at rates of four to six miles per hour for up to 22 miles, chasing down the antelope until it stops or even collapses. As biologist David

Carrier noted, twentieth-century people in Africa, Mexico, Australia, and the American Southwest still chased larger, faster animals until the beasts fatigued and could be killed with a spear or even bare hands.

Why do the animals fatigue? Consider an experiment in which two cheetahs were put on a treadmill for 15 minutes while their body temperatures were monitored with a rectal thermometer. After sprinting a little more than a mile, which is a cheetah's approximate known hunting range in the wild, their temperatures rose to 105 degrees Fahrenheit. At this high body temperature, the cheetahs stopped running, lay down on their backs, and slid down the treadmill. Every animal's endurance is limited by its ability to keep its body temperature from rising to dangerous levels during exercise.

HOW WE'RE BUILT TO MEET THE CHALLENGE

All warm-blooded animals have to maintain their temperature within a very narrow range, which is centered around 98.6 degrees Fahrenheit in humans and a slightly higher temperature in most other mammals. To keep our temperatures above the ambient temperature, we rely on the heat we generate by burning calories. But we also need to dissipate heat if the ambient temperature rises or if physical activity requires us to burn a lot of calories quickly and generate more heat than we need. And our big advantage over our natural meat sources, especially when it's hot outside, is that our ability to avoid overheating during prolonged exertion gives us more endurance than our prey.

We can dissipate some heat by inhaling the usually cooler air of the atmosphere and exchanging it for exhaled air at body temperature. And as the ambient temperature gets lower, we can passively lose more heat into the cooler air. In most circumstances, however, we eliminate the vast majority of our excess heat by sweating.

We have two types of sweat glands, *apocrine* and *eccrine*. The apocrine glands in our armpits and groins secrete a thick, cloudy, pungent sweat from their location in the hair follicles. This sweat is commonly attractive to bacteria, whose growth causes an odor. Human apocrine glands can't secrete enough sweat to cool us down significantly.

By comparison, we have approximately two million eccrine glands, which are especially concentrated in the palms of our hands, the soles of our feet, and our scalps. These glands, which can routinely produce half a quart of thin, watery sweat per hour in manual laborers, can secrete as much as 1.5–2 quarts of sweat per hour during vigorous sustained exercise such as marathon running or playing a soccer game in 95-degree temperatures. And they can produce as much as 3.5 quarts per hour in a highly acclimated person living in the tropics.

How about other mammals? Most non-primates typically have a limited number of eccrine glands in their snouts and footpads. Their predominant method for dissipating heat is to pant—by breathing more rapidly, they can exchange more of the hot air in their lungs for the cooler outside air. But panting is an inefficient way to dissipate heat, especially as it gets hot outside—and that's why cheetahs can't run for nearly as long as we can. Hoofed animals, such as horses, rely principally on their apocrine glands— that's why a sweaty horse gets so frothy, but it's also why an exercising horse heats up faster than a human.

Our unparalleled ability to sweat is the key to our endurance because it lets us maintain normal or near-normal body temperature even during vigorous exercise. Here's how it works. When our body produces sweat, it then requires energy to evaporate the water in the sweat from the surface of our skin. The energy that goes into evaporating the water is released from the body as heat, causing the body to cool off. At the human peak of about 3.5 quarts of sweat per hour—and assuming about half drips off and half evaporates from the skin surface—the evaporation of our sweat produces a cooling power of almost 1.2 kilowatts—the equivalent of about 1.6 horsepower, or about 4,000 BTUs. How much cooling power is that? It's about the same as you would get from running a standard bedroom air conditioner for 30 minutes or so, depending on its efficiency. So, at our best, with access to enough water to replace our sweat, we can cool our own bodies as well as an air conditioner can cool a bedroom! This remarkable cooling capacity allows us to perform activities that would otherwise raise our body temperature to fatal levels.

This advantage also helps explain how humans displaced Neanderthals, whose brains were as big as ours and whose bodies were stronger. Humans could walk and run farther and more quickly than the more muscular and stocky Neanderthals while requiring about 30 percent fewer calories to do so—and therefore generating less heat. And our relatively hairless bodies were better at cooling than were the hairier Neanderthals. So about 40,000 years ago, when the glaciers receded and exposed more dry land for possible habitation, we had a great survival advantage over the Neanderthals because we could travel around that open land faster and more efficiently than they could.

Water and Salt

Since the ability to sweat is so critical for humans, we also need to drink more water than other mammals. And since dehydration and low blood pressure can be so dangerous, our ancestors needed to be built to protect against these problems. And finally, since we don't have a single large water tank in which to store water for future sweating, humans have always depended on a variety of hormones—secreted by the brain, lungs, liver, kidneys, and adrenal glands—to make sure we have a slight surplus of water (and the necessary accompanying salt) spread throughout our bodies. These hormones guide the intake of water via thirst, and they guide the intake of salt through our desire for salty foods.

They also control our kidneys, which conserve salt and water when they're in short supply but eliminate them in times of excess.

We start to get dehydrated if the body's water level falls by even a modest amount. In general, trained athletes can tolerate losing two or three quarts (about four to six pounds) of water before the water deficit lowers blood pressure and can make all the cells in our body too salty for normal function. And when this happens, we won't be able to keep exercising, because we won't have enough water to make the sweat we need to maintain a normal body temperature.

Thirst originates in a specific part of the brain that's exquisitely sensitive to as little as a 1–2 percent change in the amount of body water or in the concentration of salts and other substances within the blood. Thirst can be stimulated by dehydration, regardless of whether the concentrations of salts are altered. Or it can be stimulated when the concentration of salts in the body is too high, even if total body water itself is normal. Thirst is such a carefully regulated survival trait that a pregnant mother's ovaries secrete a hormone called relaxin that maintains thirst even when she is well hydrated, because dehydration can be so deleterious to both her and her fetus.

Although salts can be composed of any combination of a base (for example, sodium, potassium, calcium, or magnesium) and an acid (for example, chloride, iodide, fluoride, or sulfate), what we usually think of as salt is table salt—sodium chloride, which is composed of one atom of sodium bound to one atom of chlorine. Each day, we need to eat enough salt to offset what we lose in our urine, sweat, and feces, so we crave salt just as we experience thirst. Because the sodium atom is lighter than chlorine, it accounts for about 40 percent of the weight of table salt. Doing the arithmetic, we find that every 2.5 grams of salt includes about 1 gram of sodium. All the body's 50–80 or so grams of sodium (depending on how big we are) comes from the dietary salt we consume.

Like Goldilocks's porridge—which couldn't be too hot or too

cold—we can't survive if our bodies' salt, water, and relative saltiness levels get too high or too low. Excessive blood levels of sodium draw water out of the body's cells and dehydrate them, essentially poisoning them, which is why drinking ocean water would kill us. At the other extreme, low levels of sodium in the blood can precipitate confusion and even seizures. Our hormones must fine-tune our water and sodium levels—accurately sensing an imbalance and fixing it as promptly as possible.

When our bodies need salt, it tastes even better to us and we want to eat more of it. But to understand the true power of our craving, consider the extreme case of a three-year-old boy cited in 1940 by medical researchers Lawson Wilkins and Curt Richter. Malfunctioning adrenal glands caused the boy to lose large amounts of salt in his urine. By age one, he licked salt from crackers and always asked for more. He loved to eat pretzels, potato chips, bacon, and pickles. He would readily eat things he otherwise didn't like, such as vegetables and even cereal, if salt were added. And when he had access to the salt shaker, he would pour salt on his plate and eat it with his fingers, consuming a teaspoon or more of pure table salt a day in addition to all that was added to his food. Perhaps not surprisingly, *salt* was one of his first words. He was ultimately admitted to a hospital for evaluation and, tragically, died when his salt intake was restricted. Because this boy's adrenal glands were abnormal, his unusual salt craving was actually an appropriate response. His sad story demonstrates how much we crave needed salt, the great lengths we will go to get it, and the fatal consequences of failing in these efforts.

But even when we don't need salt, numerous experiments show that our taste buds and overprotective survival instincts stimulate us to choose salty foods. For example, studies show that infants routinely pick saltier options by the age of four months. Makers of baby food recognize this preference and sell many products that contain half a gram of sodium per serving—nearly half a baby's daily needs in just one jar. Restaurants and processed-food manufacturers know it, too. Together they now supply about three-quarters of our daily salt intake.

If we didn't sweat and never vomited or had diarrhea, our daily salt needs would be trivial. However, sweat contains an average of about a gram of sodium per quart (less in well-trained athletes, somewhat more in the rest of us)—which is about one-third as high as the normal sodium concentration in blood and in our cells. If we sweat half a quart per hour while performing heavy manual labor for ten or so hours, we'll lose about five grams of sodium per day. Assuming most of us sweat much less than a quart per day, one gram of sodium per day would be more than enough to replace what we would likely lose in our sweat.

Modern American adults consume an average of about 3.6 grams of sodium per day, and the worldwide average is even higher—almost five grams per day. We typically eliminate most of this sodium in the urine, because our kidneys must protect us from getting too salty. But our kidneys can easily reduce this loss to far, far less if they aren't trying to eliminate excess salt. For example, the Yanomami Indians of northern Brazil and southern Venezuela have limited access to salt and can routinely limit their urinary sodium loss to about 23 milligrams per day, or about one half of 1 percent (1/200th) of our typical urinary loss.

Although a number of organs are involved in salt and water balance, our kidneys do most of the fine-tuning. Since we drink fluids and eat salt intermittently, our kidneys hold on to water and sodium in times of shortages and eliminate them in the urine in times of excess. The kidneys keep the total volume of water within a very narrow range, usually within just several percent above or below the ideal.

When blood flow to the kidneys is reduced, as it is when we're dehydrated, the kidneys avidly conserve sodium and water proportionately to balance our normal saltiness. As a result, the production of urine diminishes. In the face of modest dehydration, we can reduce the loss of water in the urine to less than half a quart per day, just enough to eliminate the waste products that would otherwise poison us.

The sensation of inadequate body water also stimulates the muscles of the arteries and veins to contract to try to maintain adequate blood pressure and increase the return of blood to the heart, despite a reduced blood volume. The heart is stimulated to pump more strongly and rapidly, so that the reduced amount of blood circulates more frequently.

As discussed in chapter 2, it's far better to be overprotected if you aren't sure what tomorrow will bring. For sodium and water, the survival benefit of overprotection is simple. If our bodies don't have enough of them, the resulting dehydration could lower our blood pressure below the threshold needed to get enough blood to our body. If blood pressure is too low, we faint or even die. By comparison, if we have a little excess sodium and water, our blood pressure is less likely to fall to dangerous levels if we sweat a lot, get diarrhea, or can't drink enough water for a while. And even if the resulting excess raises our blood pressure somewhat, this elevated blood pressure is usually well tolerated for many years. So it makes sense that our body's fine-tuning mechanisms focus mostly on making sure our water and sodium levels aren't too low rather than on worrying about whether we have a slight excess.

Over many generations, we would expect that humans with the biggest reserves of water and sodium would be the ones most likely to be able to resist dehydration, survive better, and pass that genetic advantage on to us, their descendants. If, however, you can't access enough water and salt or if you get confused or faint from dehydration, these protective mechanisms may not be sufficient to save you.

WHEN OUR SURVIVAL TRAITS ARE INADEQUATE

In late summer of 490 BCE, the Athenian army defeated the Persian army on the plains near Marathon. The victory was as definitive as it was unexpected. The Persians, who initially had a larger force on the battlefield, split their troops, sending some by boat to the port of Athens, where they mistakenly thought a civil war had weakened the city's defenses. The Athenian army decisively defeated the Persian ground troops that remained at Marathon, and the Persian naval contingent never directly attacked Athens after recognizing that the city was united and well defended.

As recounted in Plutarch's "On the Glory of the Athenians" in the first century CE, a messenger, probably named Pheidippides, was sent by the Athenian general Miltiades to inform the citizenry of Athens of their army's great victory. The distance from Marathon to Athens is about 22 miles by the hilliest route and 25 miles by a somewhat flatter route. After running to Athens in the heat of late summer, Pheidippides reportedly dropped dead of exhaustion immediately after delivering his victory message.

At first glance, it may not seem surprising that the runner of the original marathon died of exhaustion. But in reality, it's quite strange. Perhaps, as some versions of the legend suggest, this same Pheidippides had been the messenger sent from Athens to Sparta in an unsuccessful attempt to enlist the Spartans as an ally against the Persians. If so, he would have run about 150 miles each way in the days or weeks before his marathon. Or perhaps, as other versions suggest, he had already fought in the brutal battle. But regardless of how tired Pheidippides may have been at the outset of his run to Athens, Miltiades probably chose him because he was the best possible messenger, the one most likely to be able to make the solo trip through the mountains in the late summer heat to get the crucial message to the people back home.

In understanding the legend, if not the facts, at least two major questions arise. First, why did Miltiades send a messenger on foot rather than on horseback? It wasn't for a lack of horses: the animals were introduced into Greece more than 1,500 years earlier, and the ancient Olympic Games included both human races and horse races. Second, why would the best runner in the entire

Greek army collapse and die at the end of his run from Marathon to Athens, a distance that was actually at least a mile shorter than the modern marathon? Massachusetts runner Dick Hoyt, now nearly 75 years old, has completed more than 70 marathons, including more than 30 Boston Marathons, all after the age of 37. And, in each of them, he has not only run himself but also pushed a wheelchair carrying his son, Rick, who suffers from cerebral palsy.

At the beginning of a run, humans can accelerate almost as fast as horses. At 100 yards, the Olympic gold medalist Jesse Owens once beat a horse. At a mile or so, however, a horse's time will be less than two minutes, whereas the best human will clock in at slightly less than four minutes. But what about longer distances? Starting in 1980, an annual "man-versus-horse marathon" has taken place in Llanwrtyd Wells, in Wales. For the first 24 races, horse beat man, although the margins were not as great as one might think. The course was subsequently amended to include a variety of natural impediments, and in 2004 a man beat all the horses, by about two minutes.

Interestingly, both this first human victory and a subsequent one in 2007 took place in hot weather, whereas in 2011 a horse beat all the men by 17 minutes on a rainy, windy day. So on a hot day in late summer, across a terrain perhaps no smoother than that of Wales, it actually made sense for Miltiades to send his best runner on foot rather than a man on horseback.

Even before the story of Pheidippides defined an Olympic event, it also coined the term *marathon*—a protracted, seemingly endless endeavor that pushes endurance to the limit. But in the history of the human species, running the marathon is like child's play. Ultra-marathoners run about 62 miles. The Ironman Triathlon consists of a 2.4-mile swim, a 112-mile bicycle ride, and then a 26.2-mile marathon without a break. And even these events are little more than diversionary recreation compared with the distances our ancestors ran to hunt meat on the hoof.

Pheidippides's assignment wasn't a suicide mission but rather an easily achievable afternoon jog for the best person the Athenian army had to offer. But his death shows the fragility of the human body when pushed toward its maximum capacity for exercise. The ability to survive a marathon depends on our ability to maintain normal body temperature, which requires sweating, which requires adequate water and salt.

At warm outdoor temperatures, our bodies heat up rapidly because it's hard to dissipate heat quickly into the ambient air, whether by breathing or sweating. As a result, our exercise capacity falls. For example, maximum exercise is about 25 percent lower at 95 degrees Fahrenheit than at 60 degrees Fahrenheit. We also cool down faster in dry air, which can accept more of the evaporating water. In humid air, even at the same temperature, evaporation is slower, sweat stays on our skin longer, and we sweat more because we remain hot. As a result, when we exercise in humid climates, we heat up faster, the risk of dehydration is greater, and our endurance declines—maximum exercise capacity is about 30 percent lower at a relative humidity of 80 percent than it is at a relative humidity of 24 percent.

For more intense exercise, we can improve endurance by about 4 percent if (like the Kalahari Bushmen) we hydrate well before exercising, by about 6 percent if we cool our whole bodies before exercising, and by about 10 percent if we drink cold fluids while exercising. We can also exercise longer if we pour water on ourselves—that water evaporates just as our sweat does—and especially if we pour water on our heads to cool down the brain, which gets about half a degree hotter than the rest of the body. And, most remarkably, swilling menthol, which stimulates the body's cold receptors, can improve peak exercise performance by about 10 percent.

But like any adaptive mechanism, the drinking of water can be taken to an illogical extreme. Until recently, marathon runners and other avid exercisers were often advised to drink in anticipation of perspiration, or to continue to drink above and beyond their thirst so they could continue to urinate despite active exercise. But this strategy is ill advised because it can lead to a serious imbalance in the levels of water and salt. We're best governed by responding to thirst and drinking whatever we feel is necessary — essentially using the same survival mechanism that perpetuated humanity. Drinking excess water in a misguided attempt to keep urine flowing will markedly increase the risk that the blood concentration of sodium will fall below levels needed to maintain normal brain functioning. A reduction in urination during active exercise isn't abnormal — it's actually an appropriate response to stress.

Although sweat is only about one-third as salty as our blood and body cells, profuse sweating can cause a substantial loss of sodium. That's why vigorous exercisers also need to replenish their sodium losses as well. This challenge is especially critical among athletes who exercise strenuously in extreme settings — such as University of Florida football players who begin practice in the heat and humidity of late summer under conditions much like those faced by Pheidippides. In response, a team of university researchers developed the first commercially available sports drink — Gatorade, so named because the University of Florida's athletes are nicknamed Gators after their mascot, the alligator. Modern sports drinks are formulated with up to 40 percent of the sodium concentrations of sweat — enough to keep most of us in reasonable balance if we drink as much fluid as we lose.

Marathon runners typically finish a race with body temperatures of 101–103 degrees Fahrenheit and cool down rapidly afterward. On average, we can't keep running when our temperature reaches about 104 degrees Fahrenheit — about a degree lower than the limit for a cheetah. But if we continue to sweat, we cool down much faster than the cheetah after we stop. On the other hand, if we haven't drunk enough water and aren't hydrated enough to be able to sweat, our temperature will remain high even after we stop.

On average, we'll be fine if we lose up to 2 percent of our body weight as sweat, dehydrated but not sick if we lose 2–3 percent, but run the risk of *heat exhaustion* if we lose more than 3 percent. Since even experienced runners often underestimate their fluid losses by an average of about 50 percent, the importance of rehydration cannot be stressed enough.

Heat exhaustion is characterized by hot, sweaty skin, dizziness, headache, confusion, fatigue, weakness, nausea, and vomiting. The skin becomes red and warm as blood is diverted to it from vital internal organs, in the hope that the blood's proximity to the external atmosphere will help cool it and us down. This diversion of blood exacerbates dehydration and puts vital organs at even greater risk. If the dehydration lowers blood pressure enough to cause dizziness, the volume of blood returning to the heart and brain can be improved by lying down and raising the legs. People who suffer from heat exhaustion will generally respond rapidly to rest, oral rehydration with fluids, and sometimes active cooling of the skin with fans, air conditioning, or even ice. Cooling the skin also will allow blood to be redirected from the body's surface to vital organs.

Heat injury is even more severe than heat exhaustion, usually occurring with sustained body temperatures up to 104 degrees Fahrenheit. People with heat injury typically have hot, sweaty skin, and they commonly collapse from the dehydration and low blood pressure. Inadequate blood flow may injure the liver, kidneys, and intestines, and it may even lead to muscle breakdown, called rhabodomyolysis. Sufferers must be actively cooled to lower their body temperatures and reduce relative dehydration.

With *heatstroke,* body temperatures are usually sustained above 104 degrees Fahrenheit. Severe dehydration precludes the ability to continue sweating, and the skin is typically hot but dry. The combination of high internal temperature and reduced blood flow to vital organs, especially the brain, causes confusion and, ultimately, coma. People with heatstroke commonly suffer extensive damage to their livers, muscles, and kidneys. They require external cooling and generally benefit from the rapid infusion of cooled intravenous fluids to restore blood volume.

When Pheidippides ran without stopping across the foothills of Greece, the temperature probably was at least 80 degrees Fahrenheit. Unlike modern marathon races, his journey wouldn't have been attended by adoring spectators offering water and other refreshments. If he was sweating at the expected rate of up to two quarts or so per hour, his major risk was dehydration. As he became more dehydrated, he wouldn't have been able to continue sweating. Once his sweating stopped, his body temperature would have risen. If it rose too high, he would have suffered from heatstroke. Heatstroke caused by dehydration represents the principal risk for long-distance runners, especially in the summer heat— and it's almost certainly what killed Pheidippides.

In order to avoid death by dehydration, at least until they could bear and rear children, our ancestors needed their survival traits— such as thirst, the desire for salt, and the kidneys' ability to conserve water and sodium—to be overprotective. Unfortunately, even these overprotective survival traits weren't quite enough to save Pheidippides.

THE MARCH OF CIVILIZATION

If modern humans undergo a variety of stresses that make us thirsty, imagine the frequency with which early humans faced such challenges. For much of the history of humanity, water has been a critical commodity. Cities such as Jerusalem and Jericho rose at the sites of natural springs. Even more were built along rivers. The word *rival* comes from *rivalis,* which is Latin for "one using the same stream as another." People in dry habitats devised all sorts of ingenious water-collection and distribution systems. The citizens of Petra, in Jordan, which may be 9,000 years old, carved gutters into solid rock, sealed them with watertight plaster, and connected them to terra-cotta pipes that carried as much as 12 million gallons of fresh spring water daily. And one can marvel at the cisterns in Istanbul and the network of aqueducts that brought water throughout the settlements of the Roman Empire.

Acquiring salt has also been critically important throughout human history. Our prehistoric ancestors probably got enough salt from the blood and flesh of the animals they ate. But we faced a new challenge when we became a more agrarian species, because vegetarians don't get enough salt. That's why herbivores such as deer and horses find salt licks, while lions can ignore them. With the advent of agriculture, our ancestors also became more dependent on nonanimal sources of salt — whether from oceans, salt springs, or rock salt.

Salt is also an effective preservative that can keep meat and even vegetables from rotting, so it allowed food to be stored during times of plenty and used later during times of need. As noted by Mark Kurlansky in his impressive *Salt: A World History,* our ancestors developed remarkable public works and trade routes that moved salt, which commonly was more valuable than gold, around the world.

We know that as early as 6000 BCE, people in what is now modern-day Romania put salty spring water into ceramic vessels and boiled it until only the salt remained. By about 2000 BCE, salt was commonly used to preserve meat and fish. Salzburg, in Austria is, literally, "Salt City." Even the word *salary* comes from the Latin *salarium,* the name given to the money a Roman soldier paid to buy salt. The word *salad,* which means "salted" in Latin, is thought to be derived from the way Romans salted their leafy vegetables. The Romans also considered salt to be an aphrodisiac, probably because it's easier for a man to get an erection with a full bladder — which he's more likely to have if he eats extra salt and needs to drink more water to retain normal body saltiness.

As salt became more and more plentiful, its ability to preserve meat and other foods led to increasing salt consumption in industrialized societies. By the nineteenth century, before refrigeration and the widespread distribution of canned, frozen, and even fresh vegetables and meats, daily sodium intake is thought to have averaged about seven grams per day in Europe and an astounding 24 grams per day in the United States!

THEN AND NOW

When Pheidippides collapsed from heatstroke at the end of his marathon, the ancient Greeks couldn't provide him with air-conditioning, ice, or intravenous fluids. The internal injuries that resulted from the heat and dehydration led to his rapid demise.

Most modern humans rarely run marathons — especially in such conditions — but our water and salt systems can be stressed in other ways. Any of us can develop vomiting or diarrhea. During heat waves, elderly people are especially susceptible to dehydration because of the limitations of their hormonal orchestra — their thirst instinct is relatively blunted, and their hormones can't stimulate their kidneys to preserve salt and water to the extent they could have when they were younger. For most of us, however, the threats faced by Pheidippides and our Stone Age ancestors aren't a major concern in an era of bottled water, salty snacks, intravenous fluids, and air conditioners.

Furthermore, as we exercise less, we also sweat less — and need even less sodium. Our hunter-gatherer ancestors apparently did fine with about 700 milligrams of sodium per day, and human populations with very limited access to salt, such as the Yanomami of Brazil, can survive on daily sodium intakes of 400 milligrams or less. In societies such as theirs and in Papua New Guinea, where sodium intakes are below about 600 milligrams per day, blood pressures are nearly always normal and rise barely, if at all, with age.

The far higher American average daily intake of sodium comes overwhelmingly from processed foods (including breads, snacks, soups, and sauces) and restaurants—not from the salt shaker. Though Americans consume less salt now than we did in the nineteenth century, when it was a critical food preservative, we still take in far more than we need to perpetuate the species.

Simply put: a substance initially sought at great price and risk is now overabundant. A substance that used to be more expensive than gold is now added to all sorts of foods, not as a necessary preservative but rather as a taste enhancer. Something we used to struggle to get enough of is now so accessible that we must struggle not to get too much of it.

And it's not just our salt intake. To understand how the same protective traits that couldn't quite save Pheidippides in 490 BCE killed Franklin Roosevelt 2,000 years later, we'll need to understand blood pressure—what it is, its relationship to salt intake and our natural hormones, how and why it can get too high, and the damage its elevation can cause.

Understanding High Blood Pressure

Blood pressure, measured in millimeters of mercury (mmHg), is composed of two numbers. The first number, always the higher of the two, is the systolic blood pressure. Systolic blood pressure is the peak pressure generated by the heart, and it's the pressure that gets blood to the brain, kidneys, and most vital organs. It's determined by the amount of blood pumped by each heartbeat multiplied by the resistance of the blood vessels. The second number is the diastolic blood pressure, the pressure in the arteries in between heartbeats. Diastolic pressure represents a better estimate of the true resistance of the arteries to the flow of blood.

The exact blood pressure required to keep us alive varies from person to person. Simplistically, if you're otherwise healthy and not on medications that reduce blood pressure, the lower your natural blood pressure is (as long as you're awake and alert), the better off you are. But once it gets too low, it can be really dangerous.

Normal blood pressure is now defined as no higher than 120/80. A systolic blood pressure of up to 139 or a diastolic blood pressure of up to 89 is borderline elevated. A systolic blood pressure above 140 or a diastolic blood pressure above 90 is considered to be high and worthy of treatment in people under age 60, while many experts now believe it's okay for the systolic blood pressure to be as high as 150 in people over age 60 before recommending that they take medications. When we talk about high blood pressure, we're always referring to sustained, chronic elevations, not just one or two readings. Blood pressure should be measured in each arm, while sitting, at least twice after five minutes of rest.

Blood pressure varies minute by minute in response to daily activities. For example, systolic blood pressure can rise with exercise, when your heart is beating faster and more blood is pumped with each heartbeat. But it doesn't rise proportionately to the increased blood flow because the arteries to the big muscles in our legs and arms dilate in response to exercise, especially to walking and running (*isotonic exercise*), to help the muscles get the blood they need. Similarly, arteries to the intestines dilate after eating, so the intestines get enough blood and oxygen to absorb and digest the meal. The reason we're told not to swim or exercise for an hour after eating is that the competing needs of the intestines and the muscles mean that less blood may be available for one or the other, thereby causing the muscles or intestines (or both!) to cramp.

The dilatation of arteries during isotonic exercise is normally associated with a modest decrease in diastolic blood pressure. In people with high blood pressure, however, both systolic and diastolic blood pressures commonly rise with isotonic exercise. Blood pressure also increases *transiently* with exercises like weight lifting (*isometric exercise*), because the heart pumps more blood but the arteries don't relax proportionately.

People who measure their own blood pressure, whether at home or in drugstores, will notice that 5–10-point variations are common and that any kind of stress may *transiently* raise blood pressure. People whose blood pressure transiently rises above normal during stress but then returns to normal afterward have *labile* blood pressure, not true high blood pressure. Some people even have *white-coat hypertension,* which comes on when the stress of being in a doctor's office causes a truly normal blood pressure to rise briefly into the abnormal range during a doctor's or nurse's visit.

All of us suffer from periodic stress, but not all of us get transient high blood pressure with it; people with labile blood pressure often have more reactive arteries and a greater risk of developing chronic high blood pressure than the rest of us. But this risk is because of their intrinsic proclivity toward high blood pressure, not because stress per se causes chronic, *sustained* high blood pressure.

One of the great misperceptions is that we can tell if our blood pressure is elevated. But in fact, elevated blood pressure itself causes no symptoms. High blood pressure is called *hypertension,* but the condition really has nothing to do with a person being *hyper tense;* and blood pressures, on average, aren't chronically higher in people who are anxious or tense than they are in people who feel relaxed and calm.

What Causes High Blood Pressure?

If stress doesn't cause chronic, persistent high blood pressure, what does? From the time of the sphygmomanometer's general availability, after 1905 or so, the conventional wisdom was that blood pressure routinely and naturally increased with advancing age, more in some of us than in others. The belief was that all our arteries naturally stiffened and became more resistant over time, so that blood pressure rose even if the amount of blood pumped with each heartbeat remained the same.

And since rising blood pressure was nearly ubiquitous, it wasn't thought to be dangerous. Even in the late 1940s, the most widely acclaimed American textbook of cardiology defined mild, benign hypertension as a blood pressure of up to 200 systolic and 100 diastolic. The only recommended treatments were mild sedatives and, if needed, weight reduction.

We now know that this was all wrong. Normal blood pressure shouldn't be redefined with age. In the developed world, systolic blood pressure does tend to increase gradually with age — not because it's normal but rather because so many of us, over time, develop abnormal, progressive stiffening (or hardening) of the arteries, which can't relax normally to allow blood to flow through them during systole. By comparison, diastolic blood pressure tends to peak between ages 50 and 59 and may decline thereafter, because the stiffened, less elastic blood vessels don't provide the same rebound resistance in between heartbeats during diastole.

Our arteries can progressively harden in two ways. First, the thin inner layer, where the blood flows, hardens if plaques of fat, principally composed of cholesterol, are deposited on its surface. This process is known as atherosclerosis, Greek for "hard [*sclerosis*] gruel [*athero*]."

Second, all arteries have a layer of tiny muscles that can constrict or relax. When these muscles constrict, the arteries become more resistant to blood flow — so blood pressure rises. As we discussed earlier, this constriction is critical for maintaining our blood pressure when we get dehydrated. But when our blood pressure is even mildly elevated, this muscular layer constricts for a different reason — to protect our organs from being damaged by too much blood flow at an increased pressure. This process becomes a vicious cycle, because the constricted arteries compel the heart to generate higher pressures, thereby constricting the arteries even more. So once blood pressure is high, it usually gets progressively higher as we age. The higher pressures generated by this vicious cycle slowly but surely not only stiffen large arteries but also damage even the smallest branch arteries.

About 95 percent of people with high blood pressure have what is currently called *essential hypertension,* a fancy way of saying that it's the direct result of a misset sodium thermostat. A misset sodium thermostat happens because of an excess of one or more of the hormones that conserve sodium or constrict arteries to protect us from dehydration. These high hormone levels may be caused by mutations in more than 40 specific genes that are already known to play a role in regulating them. Now that we live longer and chronically eat far more sodium than we need, these overprotective thermostats, based on mutations that helped save our ancestors when they got dehydrated, have the unintended consequence of causing and progressively exacerbating high blood pressure.

The other 5 percent of cases of high blood pressure are caused by definable conditions that throw our normal water and salt systems out of balance. About 2 percent of those cases are attributable to a narrowing of the main artery to one or both kidneys. The arteries to the kidneys can be narrowed either by atherosclerosis or by a rare condition called fibromuscular dysplasia, characterized by excess buildup of arterial muscle tissue. Why is this problematic? The kidneys sense *cardiac output,* the amount of blood the heart pumps to the body each minute, as their method for detecting dehydration—for example, they know that cardiac output declines when we're dehydrated. But unfortunately, this method is as imperfect as it is simple. If there's a selective narrowing of the arteries that supply blood to the kidneys, the kidneys' estimation of cardiac output as a reflection of the body's need for salt and water can go awry. Total cardiac output could be just fine, but the restricted blood flow to the kidneys fools them into believing that it's actually diminished. As a result, the kidneys will avidly hold on to sodium and water while also releasing hormones that stimulate arteries throughout the body to constrict—just what they should do as a response to perceived dehydration.

Another 1 percent of cases occur in people in whom kidney damage itself alters the organ's ability to sense blood flow or eliminate excess sodium and water. And most of the remaining 2 percent are caused by adrenal-gland tumors that secrete excess amounts of adrenaline or other substances that constrict arteries or make us hold on to more sodium and water than we need. Unlike essential hypertension, in which the thermostats governing these various hormones are slightly misset, adrenal tumors aren't regulated by any bodily thermostats—so the high blood pressure they cause shouldn't be considered an unintended consequence of a valuable survival trait.

Our internal salt and water thermostats also explain a well-documented ethnic phenomenon. Since people are more likely to become dehydrated in hot climates than in cool ones, we might expect those of us who have lived in hot climates for many generations to be especially good at holding on to sodium and water—or better able to sustain blood pressure even when dehydrated. After all, if we can't survive without an adequate blood pressure, we must do everything we can to protect ourselves. But how does this play out in reality?

In the United States, about 30 percent of adults have high blood pressure, but the rate is about 40 percent in African American adults. Why? African Americans typically are more sensitive to any given amount of sodium than are Caucasians. As a result, the excess US sodium intake raises their blood pressures more, on average, than it raises blood pressures in Caucasians. It shouldn't be surprising that the unintended consequence of more vigorous protection against dehydration would be an especially increased risk of high blood pressure.

Excess sodium intake further revs up our body's overprotective tendency to err on the side of higher blood pressures than we need. Pretty much regardless of the cause of high blood pressure in an individual, blood pressure gets worse if you eat more salt. On average, each additional daily gram of sodium increases blood pressure by about 2.1 mmHg—and increases the likelihood of having high blood pressure by about 17 percent. Our excess sodium consumption damages our hearts, kidneys, and blood vessels—and more than six grams of sodium per day is strongly associated with an increased risk of dying. Some experts estimate that our overconsumption of sodium accounts for more than 1.5 million deaths worldwide each year.

The Damage Caused by High Blood Pressure

Blood normally flows unobstructed through the smooth inner lining of the arteries. But when blood pressure is elevated, blood flow initially becomes more rapid and can be more turbulent, just as you would observe with the flow of water through a hose when you turn up the spigot. This pressure and turbulence damage both the inner lining and the muscle layer of all arteries throughout the body, especially large arteries that carry a lot of blood and especially at branch points, where blood needs to change direction or even turn a corner. When any damaged area subsequently heals, a tiny cholesterol-containing scar may form and may narrow the artery a little bit. And as we'll learn in chapter 6, if a cholesterol-containing plaque is damaged by turbulent flow, it can fissure or rupture—and heal through the formation of a blood clot, which can further narrow the artery. If the artery is sufficiently narrowed, less blood can flow through it, and the organs beyond the narrowing suffer. And if the narrowing progresses to total blockage, the part of any organ supplied by that artery dies. In the heart, the result is a heart attack. In the brain, it's a stroke.

In addition to this narrowing from progressive damage to the internal lining of the arteries, blood pressure can get high enough literally to blow a hole in any relatively weak part of the otherwise thickened arteries—even in the aorta, the main artery that takes blood to the entire body. During a routine examination, a family doctor or ophthalmologist can see tiny thickened arteries in the retina of the eye and can detect asymptomatic leakage. Arteries in the brain are especially sensitive to rupture, perhaps because the fixed space inside the skull limits their ability to reshape in response to high blood pressure. When a brain artery springs a leak, it causes a *hemorrhagic stroke*—bleeding into the brain—as compared with the *ischemic stroke,* which is caused by a blood clot.

In the kidneys, high blood pressure damages the delicate web of arteries (the *glomerulus*) that filters toxic waste out of the blood. Each kidney has about one million glomeruli, and both kidneys together can filter about 200 quarts of water out of our blood each day. The other parts of the kidneys then remove waste products and any excess salt from the water. As noted earlier, they ultimately excrete a quart or two of water per day as urine, but they return all the rest back into the bloodstream—purified of waste and with just the right amount of salt.

But when high blood pressure damages the glomeruli, they can't function effectively, and the amount of water (with its toxic products) filtered by the kidney declines. We can do just fine if the amount filtered per day declines by 50 or even 75 percent. But when daily filtering falls below about 25 quarts, the kidneys don't have enough water to work with, and the toxic wastes accumulate in our blood and poison us—unless we're saved by a kidney dialysis machine, which filters these toxic wastes out of the blood. To make matters worse, remember what our kidneys do if they think they aren't seeing enough blood? They think we're dehydrated, and they activate hormones to conserve salt and raise blood pressure. So kidneys that are damaged by high blood pressure paradoxically do their best to raise it even higher! High blood pressure now causes more than a quarter of all cases of kidney failure in the United States.

High blood pressure also stresses the heart's ability to pump blood. Put simply, the heart muscle must grow thicker and stronger if it has to pump blood against higher resistance. But this thickening is unlike what happens when we gradually strengthen our arm muscles by lifting weights. Although high blood pressure makes the heart muscle stronger in the short term, years of high blood pressure damage it in the long term. Persistent high blood pressure eventually wears out the heart muscle, just as our arms would fatigue if we tried to lift too much weight for too long. Before the advent of effective medications, high blood pressure was a leading cause of heart failure in the United States, and it remains the leading cause in developing countries, where people are less likely to receive effective treatment.

Doctors initially thought heart failure could best be treated by using medications that make the heart muscle contract more vigorously. For example, digitalis, a medication derived from the foxglove plant, marginally improves the function of the heart muscle. So, too, do adrenaline and several other medications that work on the heart muscle. But with the exception of digitalis, whose benefits are probably real but definitely very limited, these other medications actually cause more harm than good. Why? In heart failure, the heart muscle is already being maximally stimulated by our own hormones, so yet more overstimulation doesn't help. The best way to improve heart muscle function isn't to flog it more but rather to block the vicious cycle of hormones that cause us to hold on to too much salt and water—the same hormones that cause high blood pressure. Amazingly, we now treat heart failure with the same medications we use to treat high blood pressure!

Roosevelt's High Blood Pressure

Although the residual effects of his polio left Roosevelt unable to stand without braces, the American public considered the 50-year-old New Yorker to be healthy when they first elected him president in 1932. The strength of his shoulders and arms, toughened by swimming and other activities to compensate for his paralyzed legs, made Roosevelt look robust and vigorous. Yet his apparent physical vigor belied his high blood pressure.

Medical records made public after his death show that, as early as 1931, Roosevelt's blood pressure was 140/100 mmHg—which is now considered borderline. By 1937, it was substantially elevated, at 162/98, and it continued to rise during his presidency, reaching 200/108—severe high blood pressure—by the time of the D-day invasion at Normandy, in June of 1944. By the time of the Democratic convention in the summer of 1944, Roosevelt's high blood pressure and frequent headaches made it clear that he might not survive another term. His advisers, who pressured him to select Harry Truman as his running mate for his fourth presidential campaign, knew they were probably also picking the next president of the United States.

After the election, Roosevelt's blood pressure rose even more. The last time he was seen standing in public was when he gave his brief, 500-word inaugural address in January of 1945. By February, when he, Churchill, and Stalin charted the postwar peace at the Yalta Conference, his blood pressure was off the charts, at 260/150.

Roosevelt's primary doctor during his presidency was vice admiral Ross T. McIntire, an ear, nose, and throat specialist. McIntire noted Roosevelt's deteriorating health, including weight loss and shortness of breath, and attributed it mostly to the flu and bronchitis rather than to his high blood pressure and its deleterious effects on his heart. At the request of Roosevelt's daughter, McIntire finally consulted a heart specialist, Dr. Howard G. Bruenn, who correctly diagnosed heart failure caused by high blood pressure. McIntire then prescribed a low-salt diet, which was known at the time to have some benefit in lowering blood pressure, as well as injections of mercury to inhibit the kidneys' ability to retain salt and water. He also prescribed digitalis, the best medication then available to treat heart failure.

But even the best treatments of that era were predictably ineffective. Roosevelt's high blood pressure basically remained untreated by modern standards as it indisputably rose into a range known, even then, as *malignant hypertension*. His doctors knew — or should have known, even in the 1940s — that the life expectancy of a person with malignant hypertension was no more than one or two years.

Roosevelt experienced perhaps the most devastating acute effect of high blood pressure when he sprung a leak in a blood vessel in his brain and suffered a massive, fatal stroke. Since he complained of "a terrific pain in the back of my head" before collapsing into unconsciousness, he probably had what's called a *subarachnoid hemorrhage,* in which blood also gets into the spinal fluid. His blood pressure shortly after he collapsed was 300/190.

It's tempting to think of Roosevelt as being elderly when he died during his fourth term. In reality, however, he was only 63 years old, younger than both William Henry Harrison and Zachary Taylor, nineteenth-century presidents who died of "natural causes" soon after their elections. And Roosevelt wasn't even eligible for the Social Security "old age" pension system, which was established during his presidency.

The Modern Conundrum

In the United States and other developed countries, the prevalence of high blood pressure rises dramatically with age: from about 5 percent below age 35 to about 15 percent between ages 35 and 44, about 35 percent by ages 45–54, about 50 percent by ages 55–64, about 65 percent by ages 65–74, and more than 70 percent over age 75. And its impact is enormous. High blood pressure contributes to about 50 percent of all deaths from heart attacks and stroke, and it's a key cause of heart failure and kidney failure. Overall, high blood pressure is implicated in nearly 400,000 US deaths and seven to eight million worldwide deaths each year. High blood pressure is also costly. Counting the costs of health care services and missed days of work, it costs more than $75 billion each year in the United States alone. Fortunately, the cost of using medications to reduce blood pressure is well worth the benefit, and the financial savings from averted strokes and heart attacks more than offset the costs of the medications themselves when generic medications are used.

Pheidippides died when he exceeded the limits of a remarkable set of natural adaptations that helped perpetuate our human species for 10,000 generations. Franklin Roosevelt died at age 63 because those very same overprotective survival traits were too aggressive. In fact, the perpetuation of the human species doesn't require anyone to live to be 63 — we only need to live long enough to have children and raise them through their vulnerable periods. For most of our history, we had to do so in an environment that required substantial exertion in the face of limited salt and water, not to mention epidemics of infectious diarrhea. In some parts of the developing world, these threats still dwarf concerns about high blood pressure. If the costs of having a slightly overprotective system include an increased risk of high blood pressure and strokes in our sixties, it's been a small price for the human species to pay. But now, with ready access to water, salt, and even intravenous fluids, the same traits that saved humans through millennia of dehydrating activities are a mixed blessing: the 15 percent or so of all US deaths caused by high blood pressure are many, many times higher than the number caused by dehydration.

How much sodium do we Americans really need today? I'll discuss this issue more in chapter 6, but to offset what we lose from sweat, urine, and feces, our bodies now need no more than perhaps 1.5 grams of sodium per day — less than half of the US average adult daily intake of about 3.6 grams or the worldwide average of about 5 grams that I cited earlier. So it's not surprising that the American Heart Association recommends limiting daily sodium intake to that same level (1.5 grams per day), while the World Health Organization recommends 2 grams per day and the Food and Drug Administration and the Institute of Medicine recommend about 2.3 grams daily (about two-thirds of our daily average). These quantities provide enough margin for error to ensure adequate sodium even if we sweat more than a quart of fluid per day. And if we could even get close to the low-salt habits of the Yanomami and other hunter-gatherer societies, we might be able to avoid high blood pressure altogether.

CHAPTER 4

Danger, Memory, Fear, and the Modern Epidemics of Anxiety and Depression

Just ten days before Valentine's Day in 2012, former staff sergeant Jason Pemberton, an Iraq War veteran from the army's Eighty-Second Airborne Division, shot and killed his wife, Tiffany, and then himself at their apartment in Florida. The 28-year-old Pemberton, who joined the army at age 18, and 25-year-old Tiffany had been married for about a year.

Pemberton, who had served as a forward scout — gathering and transmitting information while also serving as a sniper when needed — wasn't just any soldier. During his time in Iraq, he had been awarded three Purple Hearts.

Why do soldiers who have survived battle, often by killing others, develop post-traumatic stress disorder (PTSD) and suddenly decide to take their own lives? And even worse, why take the lives of their loved ones?

The answers aren't simple. But to understand how a former model soldier such as Jason Pemberton could snap and commit such a brutal crime, we have to understand the challenges our ancestors have faced since the Paleolithic era. They were willing to kill other people, not only in self-defense or to get food, water, or a mate but also sometimes just for the sake of honor or pride. And to avoid getting killed — by murder or other means — humans needed to be fearful. Some fears are innate and part of our reflexive behavior; others we learn from experience. Depending on the circumstances, sometimes we'll fight back, sometimes we'll flee from danger, and other times we'll be submissive and hope the bully doesn't hurt us too badly.

But the same hypervigilant, fear-driven survival mechanisms that helped our ancestors know, learn, and remember how to avoid getting killed have substantial side effects — they can cause anxiety, phobias, depression, and even suicide. And the balance of benefits to side effects has changed as the world has become safer. In modern America, an estimated one in ten adults suffers from depression, and about 40,000 people take their own lives each year. Americans now are more than twice as likely to die by suicide as they are to be murdered or to be killed in combat. Ironically, the same inborn behaviors that protected us in the Paleolithic era now kill more of us than do the challenges they were designed to counteract. And these problems may only get worse as our world continues to progress from a hostile environment with its historic challenges to a civilized society in which mortal force is no longer an appropriate response to more subtle threats.

THE SURVIVAL CHALLENGE

Prehistoric bands of hunter-gatherers, which probably averaged 30 or so related adults plus their children, foraged in a relatively organized social structure with few material assets. Even if more than one band lived in the same local geographic area, the broader neighborhood likely consisted of fewer than 200 people. These groups could serve not only as trading partners and sources of potential mates but also as hostile competitors for food and water.

Our ancestors faced a panoply of physical challenges and

potential predators, both within their tightly knit groups and beyond. If you were ostracized from your band and had to go it alone, your chances of surviving as a solitary human were very low. And even if you did survive, your genes wouldn't be perpetuated unless you stole a mate or were accepted back into a group.

When faced with danger, we commonly think of two possible courses of action — fight like a lion or flee like a mouse. However, our ancestors often faced situations in which neither would have been feasible. If you're neither strong enough to engage in a potentially successful fight nor able to escape, another alternative is to be nonthreatening and submissive, like a deer freezing in the headlights or a dog whimpering with its tail between its legs. The goal of submission is that the predator — whether a wild animal or another human — will let you live. In other words, don't fight a fight that can't be won or run a race you know you'll lose. Instead, be deferential and conciliatory in the hope that you'll be spared physical harm or at least survive to be better positioned for the future.

BACK TO THE PALEOLITHIC ERA

In 1991, the corpse of Ötzi the Iceman was discovered in the Tyrolean Alps in a melting glacier. This remarkably preserved human from about 5,000 years ago was carrying an axe, a dagger, and a quiver of arrows. Although scientists originally assumed that he had somehow stumbled while hunting and perhaps frozen to death, x-rays later revealed an arrowhead in his shoulder. Further forensic analyses showed several unhealed wounds as well as the blood of at least four people on his arrows, dagger, and clothing. We don't know exactly how he died, but he certainly didn't die peacefully. In 1996, the 9,400-year-old Kennewick Man was found along the banks of the Columbia River in Washington State. Analysis of his skeleton showed a stone projectile embedded in his pelvis, with a nearby fracture.

At first, many archaeologists thought these two skeletons were exceptions to the general rule that our ancestors were peaceful foragers, hunters, and farmers. But further evidence quickly disproved the myth of the gentle savage.

As described in Lawrence Keeley's *War Before Civilization* and expanded upon in Steven Pinker's *The Better Angels of Our Nature*, skeletal remains across multiple archaeological sites show that about 15 percent of prehistoric human hunter-gatherers died violent deaths as they fought with nature and one another for resources, mates, revenge, or pride. We don't know exactly why they killed one another at such a high rate, but we can make some educated guesses based on surviving hunter-gatherer societies in modern America, Asia, and Australia—which, remarkably, have an almost identical average rate of violent death as our Paleolithic ancestors did.

Among the Aché of Paraguay, for example, ritual club fighting among local bands routinely causes skull fractures and fatalities. Including within-band killings of infants and children, ritual fighting, and true warfare, about 60 percent of deaths among Aché children and about 40 percent among adults are caused by violence. In Papua New Guinea, 20–30 percent of deaths among adult men are attributable to murder by other adult men. About 25 percent of the adult male Yanomamis of Venezuela, who live in a state of nearly constant warfare, are killed either by other Yanomamis or in battles with other local groups, most often in conflicts over women or to get revenge rather than in fights over food. Among the Hiwi hunter-gatherers, also of Venezuela, violence accounts for about 15 percent of deaths.

A careful anthropological study revealed that none of the Hiwis were killed by predators, despite the jaguars, piranhas, and anacondas in the local environment. More than half the victims were murdered by other Hiwis, about another 25 percent were killed by other Venezuelans, and about 25 percent were killed by accidents or environmental hazards. The single leading cause of violent death was infanticide—child killing, especially of female children. The intra-Hiwi adult murders were usually attributed to fights over women, retribution by jealous husbands, or revenge for past killings. Murders by other Venezuelans were typically triggered by territorial disputes. Even among the !Kung of Botswana, often called gentle people, the murder rate is about six times higher than the peak rate ever reported in the United States and is exceeded only by two modern countries—El Salvador and Honduras, which are battlegrounds for Central American drug wars.

From these case studies, it's clear that prehistoric violent deaths were overwhelmingly caused by murder rather than by animal attacks or falling off cliffs. These statistics raise an obvious question—why have we had the urge to murder one another for 10,000 generations?

Psychologists David Buss and Joshua Duntley argue that the ability to commit murder is an evolutionary asset that's been preserved and enhanced since the Paleolithic era. Historically, the murderer, who has almost always been male, got the food, the water, and the girl, along with her reproductive potential. Murder could also be a means to maintain or enhance social status or to preserve honor—and to reap all the secondary gains related to displays of dominance. We're not here because our ancestors didn't fight but rather because when they fought, they won.

In other words, we're the descendants of the killers, not the killed. Sometimes our ancestors could survive in other ways—by fleeing, negotiating, or even becoming submissive—and all these traits are perpetuated somewhere in our DNA. But one thing is clear. The meek didn't dominate in the Paleolithic era, and the descendants of the meek have not, at least yet, inherited the earth.

But the problem is that the physical ability to commit murder and the defensive traits needed to avoid it become not only far less important but also detrimental in a developed society with laws and law enforcement agencies. We're now more likely to compete in sport, in business, and for social status rather than use physical violence in life-or-death battles. As a result, many of our defense mechanisms may be ill suited to the real challenges we now face.

HOW WE'RE BUILT TO MEET THE CHALLENGE

Killers are obviously much more likely to pass on their genes than their victims are. As we saw in chapter 1, a genetic mutation that provides just a 10 percent survival advantage can spread to nearly 100 percent of the population in 150 generations, or about 3,000 years. We don't have an identifiable "murder gene," but more than 90 percent of men and more than 80 percent of women will, on close questioning, report explicitly homicidal thoughts. Nearly every person is willing to kill another person if the situation is dire enough — such as self-defense or defense of their family.

At least 85 percent of murders are committed by men, and about 80 percent of the fatalities are men as well. In developed countries such as the United States, the predilection of men to commit murder has sometimes been blamed on cultural influences, ranging from the prevalence of toy guns in childhood to violent movies and video games. But men commit a similar percentage of murders in all cultures, independent of toy stores, movies, and television.

From an evolutionary perspective, men may kill other men in order to obtain resources, reduce threats from rivals, defend against attacks, or establish superiority, thereby discouraging future attacks. Murder is more than just violent behavior taken to an extreme: it also serves a clear evolutionary purpose. Many of the benefits accruing to the murderer wouldn't be realized if the victim could lick his wounds and survive to fight another day. The murder victim loses his own life and the opportunity to have future children. In addition, the victim's existing children will find it more difficult to compete successfully with the offspring of the murderer and ultimately, if not immediately, be less likely to survive.

Successful aggression increases resources, mating opportunities, and the likelihood of passing on your genes. For example, the Mongols perpetrated some of the biggest mass slaughters of all time. Was it worth it? Well, it turns out that a remarkable 8 percent of men living in what used to be the Mongol Empire have the same Y chromosome, which can be traced back to the conquest by Genghis Khan and his sons. So in terms of not only gaining territory but also maximizing subsequent genetic influence, Genghis and the Mongols actually got what they wanted.

Men also kill women. More than one-third of murdered women worldwide are killed by an intimate partner, particularly when the woman is in her prime reproductive years—and particularly during the interval beginning after the murderous mate or ex-mate believes she has been irretrievably lost and continuing until she enters a definitive, long-term relationship with another man whose genes she might perpetuate and who might protect her. By comparison, nonfatal violence against an intimate partner—which frequently is a precursor to homicide—is designed to discourage or punish infidelity while retaining the mate. Put simply, it's counterproductive for a man to kill a mate if he thinks that nonfatal violence could guarantee her faithfulness.

Another form of murder is infanticide, practiced principally by women who are reluctant to divert scarce resources to a weak child and away from others in whom substantial efforts have already been invested. When men kill children, it's overwhelmingly because they know or fear that the child isn't truly theirs, and they don't want to invest any energy in perpetuating someone else's genes. This phenomenon also explains the *Cinderella effect*— why a child is about 100 times more likely to be killed by a stepparent than by a biological parent.

From an evolutionary perspective, sexual competition is the reason young men tend to fight with other young men for status and honor. It's also why they use violence, even lethal violence, against a mate suspected of infidelity. Given the limited amount of female reproductive capacity in any group, men are competing with one another for reproductive opportunities in a zero-sum game. A man can have many more surviving children than a woman, but he's also more likely than a woman to have none at all.

Much of a man's aggressiveness is driven by testosterone — the male sex steroid that builds muscles and must be distinguished from the stress steroid, cortisol. Studies show that when a man sees an attractive young woman or engages in a physical competition, his testosterone level will rise acutely, and he will quickly become more prone to risk taking, whether it really helps him win or he just wants to show off. Testosterone levels tend to decline in married men, probably because they no longer have to compete as aggressively to find a woman and because they perform better as mates and fathers if they calm down a bit. So it's not surprising that one study of delinquent Boston teenagers showed that about 75 percent of those who remained bachelors committed crimes as adults, compared with only about one-third of those who got married.

But reduced testosterone levels and domestic responsibilities can be problematic in other ways. For example, male professional tennis players don't do nearly as well in the year after they get married as they did in the year before, whereas the performance of their unmarried colleagues of the same age doesn't decline.

Being related — kinship — generally increases the likelihood of forming alliances and decreases the likelihood of violence against a member of the clan. In one study in Miami, about 30 percent of conspiring murderers were genetic relatives, but only 2 percent of victims were killed by genetic relatives. Even within an individual household, a murderer is more than ten times as likely to kill someone who isn't genetically related than to kill someone who is.

Ever since prehistoric times, murder could escalate into war or even genocide, in which all the losers, often including women and children, are killed and their dead bodies mutilated or even eaten. But lest we think of these as totally uncivilized and immoral acts, consider the biblical passage in Numbers 31. After the Israelites defeated the Midianites, they killed the men, plundered the villages, and took the women and children hostage. But that wasn't enough. Moses, presumably following directions from above, instructed the Israelites to kill all the male children and all the nonvirgin women. What better way to assure that the Israelites' male genes, but none of the Midianites' male genes, could be perpetuated?

Memory and Fear

Just as the ability to commit murder confers a survival advantage, so, too, does anything that can substantially reduce your risk of getting killed. It's like an arms race between two interrelated yet competing instincts — how to kill when it will give us and our genes a survival advantage and how to use our brains to avoid getting killed for the same reason. We can't identify the precise genetic mutations that drive either of these behaviors, nor can we quantify the exact rapidity of their spread, but surely these survival traits rank right up there with appropriately colored skin as our ancestors evolved to survive.

Our brain represents only about 2 percent of our weight, but it consumes about 17 percent of our calories. Furthermore, biologists estimate that about half our genes influence how our brain functions — a percentage far higher than for any other organ. As a result, any random mutation is more likely to affect brain function — for better or worse — than it is the function of any other part of our bodies. The human brain is at least three times bigger (relative to our body size) than the brains of other animals (relative to their body size). The difference is almost all in what's called the *neocortex,* the part of the brain that becomes progressively larger in mammals, primates, and especially humans, while the more rudimentary parts of the brain remain pretty much constant for body size across the animal kingdom.

The neocortex is important not only for memory but also for what's often called emotional intelligence, which is the ability to use our memories to navigate challenging social situations. One example is the appropriate use of learned fear to avoid getting killed.

Our understanding of memory is still relatively rudimentary, but studies such as those pioneered by my Nobel Prize–winning Columbia University colleague Eric Kandel are beginning to solve the puzzle. We now know that memories are initially recorded deep in our temporal lobes, the parts of the brain that are located on the two sides of the forehead. Memories are then filed away in other parts of the brain, with the hippocampus in our temporal lobes serving not only as the file clerk but also as the search engine to retrieve memories when needed. It's even possible to create totally false memories by artificially stimulating the relevant memory-coding part of the brain. And when this part of the brain was removed to cure refractory seizures in a man named Henry Molaison in 1953, he became unable to remember anything for more than about 30 seconds and served as the real-life male counterpart to the female role played by actress Drew Barrymore in the movie *50 First Dates*.

When memories are retrieved, they aren't perfect—that's why people remember things differently and argue with one another about exactly what happened. Whenever we recall a memory, it may be strengthened or questioned for accuracy in our own minds or even modified because of intervening events. But regardless of their flaws, memories of scary or dangerous situations stick with us, and our resulting fear influences how we'll strive to avoid or deal with such situations in the future.

At the other extreme are things we never *learned* but intrinsically *know*. That's why infants *know* to have separation anxiety at about the same time they learn how to crawl but have not yet *learned* to be afraid of cars. We have a robust, innate, hardwired *fear module* that is based on evolutionary survival traits and is largely independent of conscious behavioral control. Toddlers fear snakes and spiders, but they don't fear the guns and electrical outlets that really are much more common and serious modern threats. Our hardwired fears are so strong that our heart rates will rise perceptibly if we're shown pictures of scary things such as snakes and angry faces—even if the visual image is as brief as several hundredths of a second and we can't consciously remember seeing it.

In between conscious learning on the one hand and intrinsic hardwiring on the other is an interesting phenomenon called *conditioning,* whereby memories of anything that's associated with a threat of physical harm can stimulate fear. The principle is the same as the classic Pavlovian experiment that paired giving food to a dog with the ringing of a bell; after such training, the dog would salivate in response to the sound alone. Consider the following experiment. You're hooked up to electrodes and shown a series of pictures. Every time you're shown a picture of a flower, which you normally would find pleasant, you're given an electric shock. It doesn't take long to learn to learn to fear flowers. This learning can be reversed if you're then repeatedly shown the same flowers but not shocked. And it's not just the learned fear of flowers that will disappear. Even fear of modern threats, such as electrical outlets and guns, can be diminished or extinguished in experiments in which shocks are randomly administered to a person who is viewing a range of pictures.

But if the same type of shock is paired with pictures of snakes or spiders, the fear response is more intense, and it can't be eradicated by dissociating it from a predictable painful shock. These experiments prove that some fears are hardwired whereas others are learned—whether consciously or whether by subconscious conditioning.

Children are usually not naturally afraid of fire, but they learn not to put their hands in a flame, either because they believe what they've been taught or because they've previously been burned.

But if all fear had to be learned only by personally experiencing and consciously learning from danger, most of our ancestors would never have survived.

How Fear Affects Behavior

One major advantage of fear is that it can help you avoid dangerous situations. And if your fears help you avoid being attacked, you're safe, at least for the time being. But sometimes, danger may be unavoidable. And when our ancestors were attacked by aggressive rivals despite their fearful precautions, their instincts needed to zip into action to protect them. Even today, the same defensive traits that kept our ancestors alive are part of our psychology.

The psychiatrist H. Stefan Bracha describes six ways to defend against danger or an acute attack: freeze, faint, flee, fight, be submissive, or play dead. Each of these defenses can be remarkably useful. For example, fear of heights induces paralysis (freezing), so you won't step off the edge. Fear of blood makes you faint — and when you're bleeding you're better off lying down to preserve blood flow to your brain. If you're confronted by a moose, you should run away as fast as you can. If you're faced with a predatory mountain lion, you should stand tall, make noise, throw rocks, and prepare to fight for your life. By comparison, if you find yourself face-to-face with a wolf, it's best to avoid eye contact and assume a submissive position. For bears, it's somewhere in between — don't run, but do make a lot of noise to try to scare off the bear, then lie down and play dead if the bear is undeterred. And if this isn't complicated enough, these threats are just a small sample of the many our ancestors had to deal with — sometimes by knowing innately, sometimes by being taught by their parents, and sometimes by learning firsthand.

On average, the downsides of running away from danger are much less dire than the risks of the danger itself. As psychiatrist Randolph Nesse of the University of Michigan argues, if our ancestors needed to burn 200 calories to run away from a predator—who otherwise would injure them so they would forage 20,000 fewer calories over the next two weeks or so—they should have run whenever the probability of an attack was above 1 percent. He likens the fact that it's worth it to flee 100 times to avoid one attack to the *smoke detector principle*—it's better to put up with the aggravation and inconvenience of many false alarms than to have your house burn down. And if we're now willing to have lots of false alarms to keep our houses from burning down, our ancestors certainly were willing to have even more false alarms to keep from getting killed.

But repetitive false alarms take an increasing toll on the body and can even lead us to stop responding. This phenomenon is true for any stress. Stress directly stimulates the brain's pituitary gland to release hormones that then tell our adrenal glands, located near our kidneys, to release adrenaline and cortisol. The burst of adrenaline raises our pulse, blood pressure, and awareness—much like a fire alarm—so we can quickly fight or flee. The natural release of physiological cortisol makes us more alert, helps us preserve salt, and increases our ability to fight infections. The majority of us don't suffer from an excess of these stress hormones because our fear and stress decline as we learn that most alarms are false. For others, however, when stress goes up it causes its own problems. For example, mice—like humans—can quickly learn the connection between a specific tone and an electric shock. But if they're premedicated with cortisol, they'll fear not only the tone to which they've been trained but also other sounds similar to it.

Too much cortisol can damage cells throughout the body, including our brain cells. Studies by my Columbia colleague Andrew Marks and others show that stress can damage a calcium regulator in our cells—whether heart, muscle, or brain cells—so they leak calcium, just as a damaged car engine leaks oil. And in the brain, where this same calcium regulator is critical for our basic flight-or-fight response, stress-induced damage leads to problems with learning and memory. Leaky calcium explains what we all know: when we're stressed, we don't think as clearly and we're more likely to have bad judgment—which will often further increase stress in a vicious cycle.

Insights into how stress damages cells come from studies performed by Nobel Prize–winning biologist Elizabeth Blackburn. In each of our cells, the ends of our DNA have thousands of non-coding bases called telomeres. As our cells divide and replicate, the telomeres get shorter—until they become so short that the cell can't divide or function normally. Blackburn and her colleagues compared the lengths of telomeres in mothers of healthy children with those in mothers of chronically ill children for whom the mother had been the main caretaker. They found a fascinating correlation—the longer a mother had served as the principal caretaker of her ill child, the shorter her telomeres were. In the most stressed mothers, the degree of telomere shortening was equivalent to what typically would be found with a decade or more of aging. When Blackburn and her colleagues went back to study possible reasons in the laboratory, they found that cortisol—that same stress hormone we've been talking about—reduces the activity of the enzyme that's responsible for maintaining the length of our telomeres.

Our cortisol levels, calcium regulation, and telomeres show the downsides of our response to stress and fear. Although some stress and fear are critical to staying alive, too much of them or an overreaction to them causes detectable malfunctions in the ways our bodies work and results in maladaptation in the ways we think and behave.

Going Too Far

For every key response to imminent or actual threats, there's a corresponding emotion that anticipates the threat, even though it may never occur. For example, on safari in Africa, you should be fearful if you're actually charged by a wild elephant. You might also worry because you anticipate the possibility of being charged by a wild elephant on tomorrow's safari, and that concern would be perfectly reasonable and adaptive if you accurately estimate the likelihood of such an attack. But if you substantially overestimate the risk of an elephant attack and get so nervous you can't enjoy the safari, then you've developed a maladaptive *anxiety*.

In this same context, modern anxieties can be thought of as exaggerated versions of many of the same behavioral adaptations that helped save our ancestors—only now they're doing more harm than good. A little excitement and anxiety can improve your ability to perform tasks for which you are well prepared and have the skills to do. But with too much anxiety and pressure, performance deteriorates. An appropriate level of shyness and reticence helped protect our ancestors from making enemies and being ostracized from their group. But now these same tendencies cause about 7 percent of Americans to have *social anxiety disorder*—a marked fear of embarrassment or humiliation, which can be manifested as anxiety about speaking or performing in public, difficulty making friends, and even trouble holding a conversation.

Nonspecific hypervigilance and even occasional panic may have been critical when animals, predators, and powerful and potentially homicidal humans were constant threats. Now, however, about 3 percent of American adults report *generalized anxiety*—difficulty functioning because of excessive worry about everyday challenges such as family and job responsibilities, finances, and the health and safety of themselves or family members. In about 2 percent of Americans, this out-of-proportion worry and anxiety precipitates *panic attacks*—intense fear accompanied by psychosomatic physical complaints such as heart palpitations, trembling, dizziness, or chest or abdominal pain.

Even common *phobias,* which are anxieties out of proportion to specific threats, also are directly related to historic dangers. For example, acrophobics are afraid of falling, just as our ancestors had to be careful as they traversed mountains and cliffs. And agoraphobics are afraid of open spaces, in which our ancestors couldn't hide from potential predators.

In the same way that as we can suffer maladaptive overreactions and side effects because we overestimate the likelihood of potential danger, we also can overreact to dangers we actually do face. For example, submissiveness sometimes may be the best survival strategy when faced with an impossible situation or an unwinnable challenge. However, the resulting side effect of low self-esteem can lead to overreactive social withdrawal and sadness.

Sadness was a protective trait in the Paleolithic era, when it was part of the submissiveness that kept a person from being killed by a stronger adversary. But in the modern world, sadness is usually precipitated by a variety of losses unrelated to a physical confrontation — the loss of social fulfillment, prestige, possessions, a mate, and your own health are but a few examples. In most people, sadness is reactive and often transient. But when sadness — which can range from relatively mild to severe — persists for more than about two weeks, it begins to transition into what we call *depression*. And in modern times, depression has become the most problematic side effect of our intrinsic psychological survival traits.

Just as anxiety is characterized by more fear than you need as well as fear when you don't need it, depression is characterized by exaggerated and persistent sadness — more sadness than is typical or helpful. It's easy to understand the potential prehistoric survival benefits of worrying and transient submissiveness — and even anxiety — but it's harder to fathom why persistent sadness is useful. After all, our Declaration of Independence put the pursuit of happiness right up there with life and liberty as examples of our inalienable rights.

The psychiatrist Paul Keedwell provides a perspective in his book *How Sadness Survived: The Evolutionary Basis of Depression*. As he notes, some of us are more susceptible to depression than others, but we all can become at least somewhat depressed if things get bleak enough. In this context, he argues that, on average, our ability to be sad and get depressed has probably conveyed some survival advantage—or, at the very least, has not been disadvantageous—over thousands of generations. He and other evolutionary psychologists emphasize that sadness is a hardwired mental response to losses and defeats. No one likes being sad, so sadness serves as an unpleasant trigger telling us to find alternative ways to pursue our inalienable right to happiness. Sad people involute, save energy, avoid interactions, and become self-reflective. Sadness is a wake-up call to change course—whether that means finding different strategies to pursue your original goals or substituting more reasonable and achievable goals.

But then why depression? As Keedwell emphasizes, depression ensues because of a persistent and protracted inability to achieve goals that can make us happy or to substitute more achievable ones. And when true depression develops, it's a maladaptive overresponse. Severely depressed, melancholic people stop eating, have trouble sleeping, and isolate themselves—sure ways to reduce their viability in a foraging environment. In the Paleolithic era, melancholy may have been one of the ways that weak individuals, whose dependency could slow down the entire band, were self-eliminated. For example, major depression is most common in adolescents and in the elderly—times of life when hunter-gatherers consumed more calories than they foraged.

What are the key goals that, if not achieved, may be the most likely to trigger sadness and even depression? Pretty simple—innate, hardwired human goals since the Paleolithic era, such as having enough food and water, keeping physically safe, maintaining positive human interactions, and finding a mate. In the modern world, these goals may translate into having a nice home, a loving family, healthy children, financial success, social status, and even fame. And to achieve these goals, we commonly need to have enough self-confidence to compete with others who want the same things.

In an attempt to avoid physical or other types of harm, depressed people consciously or unconsciously choose submissiveness over competition and shift from fighting to a conciliatory acceptance of failure. A depressed person appropriately disengages from tasks that are unlikely to be productive, stops trying to reach unreachable and potentially dangerous goals, and attempts to deal with self-defined disappointment and failure. If sadness and even mild depression are short-lived, they serve as a transition phase, during which a person has time to find more realistic strategies that may have a better chance of success. Rather than being killed or ostracized, you readjust and recharge your batteries, so you can compete on more favorable terms, whether in your present environment or somewhere else, at a later time. This transition period was presumably the evolutionary benefit of a time-limited depressive reaction that gave people an opportunity to step back, ruminate, and preserve energy for future success.

Depression is biochemically linked to low brain serotonin levels — a chemical imbalance that can be triggered in almost any of us by sufficiently severe life events but undoubtedly is more likely to appear in those of us with an inherited susceptibility. For example, when male primates lose a battle, suffer a decline in social status, and become subordinate to the alpha male, their serotonin levels decline. Not surprisingly, we can treat the relative deficiency in brain serotonin in human depression with medications such as selective serotonin reuptake inhibitors (SSRIs), which interfere with the degradation of serotonin and leave more of it available to bathe our brain cells and to blunt or reverse depressive symptoms.

Recently, scientists discovered that brain cells rely on another chemical, gamma-aminobutyric acid (GABA), to learn avoidance after social defeat. For example, when mice are exposed to a trained bully mouse, the mice who cower in defeat are the ones whose GABA-responsive brain cells go into overdrive. As we'll discover later, in chapter 8, medications that block GABA in the brain may be a future option for treating depression.

Why are women at least twice as likely to suffer from depression as men? We don't know for sure, but throughout our existence, women have been dependent on physically larger, dominant men, and they run the risk of serious physical harm should they be perceived as being disobedient, let alone unfaithful. Only men who lost a competition needed to be submissive, but even the women who succeeded and won their prize—the alpha male—historically had to balance their success with a healthy dose of submissiveness.

To the extent that memory-based fear protects us from danger but doesn't drive us crazy, it's appropriate and useful. But just as too much stress causes a mouse's brain to overreact, the same reaction can happen in the stressed brains of people with PTSD, in whom everyday activities can trigger memories that cause overblown fear. People who suffer from PTSD experience disturbing flashbacks, in which their memories repeatedly replay the traumatic event. In response to these continual fears, they typically become overaggressive toward anyone whom they perceive as threatening. And like embattled people who feel they have little to lose, they often become reckless or even self-destructive.

Although we tend to associate PTSD with soldiers and survivors of war or genocide, the condition can be triggered by nearly any physical or psychological trauma. The biochemical basis is uncertain, but one clue comes from studies of the brain's cannabinoid receptors—the sites where the active ingredient in marijuana binds to brain cells and initiates a process that decreases anxiety. A specific genetic mutation, found in about 20 percent of people of European descent but as many as 45 percent of some Nigerians, lowers the activity of an enzyme that inactivates anandamide, which is a natural brain substance that mimics the effects of marijuana and has been euphemistically called a bliss hormone. In both mice and humans, this anandamide-enhancing mutation reduces learned fear and the associated anxiety; it even helps extinguish preexisting fears. And perhaps not surprisingly, people with this mutation are much, much less likely to become dependent on marijuana. People with PTSD also tend to have low levels of gastrin-releasing peptide (GRP), which normally controls the fear response. Specific parts of the brain, including the memory-retrieving hippocampus, are reduced in size in people with PTSD. Although we don't yet know whether these changes in brain structure are causes or results of PTSD, they provide compelling evidence that PTSD is linked to real alterations in the brain's physiology and even anatomy.

Defeat, failure, and fear can also lead to other maladaptive behaviors that help reduce psychic pain. The well-known aphorism "drink your sorrows away" probably has a very basic explanation, which can be demonstrated in organisms as simple as fruit flies. A male fruit fly with access to welcoming females doesn't really care whether his food is or isn't fortified with alcohol. But when a male fruit fly is rejected by potential female mates, he'll strongly prefer the alcohol-laced food. Male fruit flies may not be able to drink their sorrows away, but there appears to be a reward or pleasure center in the fruit fly brain that really likes sex but finds alcohol to be a good consolation prize. In many depressed people, alcohol and other addictions complicate the situation, dull good decision making, and increase the likelihood of self-destruction and suicide.

Since our protective instincts are hardwired into our DNA, it's not surprising that essentially all our overprotective tendencies — phobias, anxieties, depression, and the like — tend to run in families. If one identical twin is depressed, there's a 40 percent chance that the other twin will be depressed as well; in fraternal twins, there's about a 20 percent chance that the other twin in the pair will be depressed. But because 60 percent of people with a depressed identical twin don't get depressed themselves, it's obvious that no inherited mutation provides a smoking gun to explain exactly who will get depressed—unlike the mutations that precisely predict lactose tolerance.

Data also suggest that stressful events cause epigenetic tagging (more about this in chapter 6), which modifies our DNA and thereby alters the brain's reaction to future stress. The same is true for other specific psychiatric diagnoses. For example, several psychiatric disorders are linked to genes that regulate calcium, but the clinical manifestations in a given individual depend on life experiences, perhaps also mediated by acquired epigenetic changes in the DNA.

Even as scientists search for precise genetic and environmental triggers for anxiety, fear, and depression, these behaviors make sense from an evolutionary perspective. Your anxieties help you anticipate danger, and your fears help you avoid danger. But when faced with an unbeatable aggressor or an impossible-to-win situation, you can protect yourself, at least temporarily, by becoming submissive, sad, and even temporarily depressed. And if a prior experience was especially stressful, your painful memories will serve as a constant reminder never to let yourself get in that situation again.

WHEN OUR SURVIVAL TRAITS ARE INADEQUATE

If all our ancestors were anxious or submissive, they never would have had the courage to explore and populate the earth. At the other extreme, however, people with too little fear were more likely to fall off cliffs, get killed by wild animals, go injudiciously into battle, and die before procreating. Our ancestors needed to walk a thin line: they had be willing to take chances yet also be cautious, vigilant, and responsive enough to survive.

To illustrate what happens when fear is insufficient, consider an experiment conducted on three groups of guppies—those who curiously inspected a potential predator on the other side of a fully protective partition, those who had no interest, and those whose responses were somewhere in between these two extremes. When a bass subsequently was put into their aquarium for 60 hours, all the curious guppies were eaten, compared with 60 percent of the uninterested group and 85 percent of the in-between group. Obviously we need some brave folk, but there are situations in which curiosity can kill more than just the cat.

The dangers of inappropriate fearlessness were demonstrated, for example, by general George Custer. In June of 1876, Custer — who had graduated last in his class at West Point — led no more than 500 cavalry and supporting troops in an attack on a Lakota and Cheyenne force generally estimated to be at least seven times larger. His lack of appropriate fear quickly resulted in the slaughter of all his troops. Oh, and by the way, most of the dead were stripped naked, scalped, and mutilated by the victors.

Although Custer took his fearlessness too far for his own health, we've always needed some less fearful ancestors to exhibit bravery and win battles. After all, if 100 percent of our ancestors were overly fearful, every one of us would probably still be foraging in Africa. Some of our successful ancestors had to be willing to take chances.

In our brains, five different receptors sense the presence of dopamine, one of the major chemicals that regulate brain activity. A specific mutation that alters the activity of the fourth receptor, called DRD4, is associated with novelty seeking, risk taking, and impulsiveness. This mutation even helps explain relative risk taking among skiers and snowboarders, people who already take more risks than average. A different mutation in this gene also is more common in people who live past the age of 90, perhaps because it encourages people to stay active and engage in the type of social and physical activities that promote longevity.

Unfortunately, mutations in this gene also have downsides. The mutation that increases life expectancy is significantly more common in children with attention-deficit/hyperactivity disorder (ADHD), especially among children who receive insufficiently sensitive early maternal care or spend a lot of time in child care. And this same mutation is associated with low educational achievement as well as high rates of infidelity and promiscuity.

But most important from an evolutionary perspective, scientists estimate that these *DRD4* mutations first appeared about 40,000 years ago—at about the same time that our ancestors made their definitive move out of Africa and began exploring the rest of the world. Mutations in *DRD4* are now found in about 20 percent of the overall human population, but they're increasingly common among people whose ancestors migrated the farthest from Africa.

It's probably good for some of us to be risk takers and some of us to be cautious. That would explain why scientists believe the *DRD4* mutations have reached an equilibrium—the 20 percent worldwide prevalence may be a desirable societal balance between risk-seeking people and their more prudent colleagues.

Independent of mutations in our *DRD4* genes, all of us tend to be brave and be willing to risk more if we think there's less to lose. That's why people with PTSD often become inappropriately tolerant of risk. It's also why American men who come from disadvantaged neighborhoods in which life expectancy is short are more likely to commit violence and murder than those who come from affluent neighborhoods. A person with little hope is more likely to risk getting killed or incarcerated in order to compete for prestige, money, street credibility, and the resulting likelihood of passing on his genes—since none would get passed on if he were unwilling to fight.

Competition, aggression, defense, memory, fear, and the emotions they elicit have always been a double-edged sword. Our Paleolithic ancestors couldn't survive if they were either too risky or too hesitant. But each of them somehow either won the battles, escaped the battlefield, or found a way to submit to the victor without getting killed—otherwise someone else's descendants would be here today.

THE MARCH OF CIVILIZATION

When our ancestors first developed agriculture about 10,000 years ago, cooperation became even more beneficial for the entire community, whether it involved watching over animals or creating irrigation systems. But there was also a lot more to fight over. Inequalities grew, polygamy became more common, and competition for resources increased. Physically dominant men, defined by their musculature and success in battle — think testosterone — had sexual relations with more women and, like Genghis and his sons, fathered more surviving children. As a result, the number of violent deaths rose to about 25 percent of all deaths, substantially higher than the 15 percent or so seen before the advent of agriculture.

Gradually, however, our ancestors formed communities with common defenses, rules, and even assessments. For example, about 6,000 years ago, British farmers constructed palisades and dug protective ditches around their settlements. Although archaeologists first thought the ditches, which were littered with human bones, were burial grounds, subsequent excavation unearthed thousands of arrowheads on the palisades and particularly around the gates. The interpretation is obvious — the ditches and palisades provided critical protection, and invaders died trying to breach them.

As settlements coalesced into cities, states, and empires, the rate of violent death declined. In Europe, violence began to decline by the eleventh century, as the culture of revenge gradually subsided and a culture of honor rose. During the reign of Henry I of England, from 1100 to 1135, homicide became a crime against the state rather than against a victim's family. The end of the Middle Ages saw the rise of an economic system that increasingly depended on commerce rather than just on land. Fighting over land is the ultimate zero-sum game — there's only so much of it, and either you have it or I have it. By comparison, trade and commerce can increase for everyone's benefit.

Interestingly, the gradual domestication of the human species, starting with cooperative bands in the Paleolithic era and continuing with the development of agricultural communities, was accompanied by objective anatomic changes. Over the past 80,000 years, the bony ridges over our eyebrows have shrunk and our faces have gotten smaller — especially among men, whose faces are now more similar to female faces than they were in the early Paleolithic era. The explanation appears to be lower testosterone levels — and gradual natural selection for men whose lower testosterone levels resulted in less aggressive behavior and a better ability to coexist without getting killed.

But even in the midst of an evolutionary trend toward sociability, we still must recognize that violence and murder can spike quickly whenever civilization breaks down or societal pressures change. For example, the Semai of Malaysia eschewed violence, probably because of a long history of being defeated by neighboring Malays, who are both more warlike and more numerous. In the 1950s, the British recruited the Semai as counterinsurgency scouts against Communist guerrillas. But after the guerrillas killed some of the Semai, their surviving colleagues became committed warriors who avidly sought fatal revenge. When the conflict ended, the Semai went home and returned to nonviolence. The Semai experience shows that even the most peaceful people will become extremely violent when provoked by revenge or the need to save themselves or their families.

Above and beyond murder rates, warfare and genocide also contribute to violent deaths. For example, recent genocides in Bosnia, Kosovo, Indonesia, Rwanda, and the Democratic Republic of the Congo have killed millions. In Bangladesh alone, the Pakistani army may have killed as many as three million people. But all the wars and genocides that made the twentieth century perhaps the most violent of the past 500 years — an estimated 180 million people killed—account for only about 3 percent of all human deaths during that period, a far lower percentage than were killed in precivilized societies.

In civilized states, a decline in violence always begins in the elite and then the middle class, who rely on manners, rules, and a formal judicial system instead of personal revenge. People of low socioeconomic status—perhaps because they have less trust in the system—are more likely to use violence rather than seek legal recourse, to cite morality as justification for revenge, and to perpetuate a cycle of violence. The famous feud between the Hatfields of West Virginia and the McCoys of Kentucky, who were families of hunters and subsistence farmers, is a good and relatively recent example of this phenomenon. But even this famous feud, with its protracted fighting and more than a dozen deaths, was minor compared with the battles of yore.

Societal norms changed more slowly in regard to a man's violence toward an unfaithful mate. In some early societies, sexual relations between a married woman and a man who was not her husband constituted a crime, akin to if the two had conspired to steal the husband's possessions. In many hunter-gatherer societies, a cuckolded man had the right to kill his wife as well as the man with whom she cheated, and the same principle was followed by the ancient Greeks and even in early English common law. Texas considered such homicide justifiable until 1974.

As a species, we have increasingly relied on laws and societal pressures to reduce violence. On a personal level, we have the unique human capacity to use memory and emotional intelligence to find nonviolent solutions through accommodation or negotiation. Consider, for example, Lewis and Clark, who were sent by president Thomas Jefferson from a civilized state to explore terrain controlled by some of history's most successful hunter-gatherers. Along the way, they interacted with more than 20 indigenous populations and faced hazards ranging from physical challenges to the hardships of winter to the need to depend upon people who had no particular incentive to help them. They used their memories, their vigilance, and their judgment to decide when to flee danger, when to fight an adversary, and when and how to negotiate their way through a difficult situation. Ultimately, they experienced only one fatal fight.

(handwritten annotation: AND SAKAJAWEA!)

In the modern world, if we agree that you won't kill me and I won't kill you, perhaps both of us can perpetuate our genes. That social progress has markedly reduced the rates of violence and murder, but it hasn't eliminated competition for success, prestige, or mates. And the same hardwired instincts of fear, anxiety, sadness, depression, and even PTSD still afflict those of us who think we've lost these modern competitions.

THEN AND NOW

As progressive civilization and commerce rewarded those who could thrive in cities without killing one another, violent deaths continued to decline. The current homicide rate in Great Britain is less than 5 percent of what it was back in the thirteenth century, and the murder rate throughout western Europe is well below 1 percent of what it was in the precivilized world. The median worldwide homicide rate is now 60 per million people per year, down by almost 50 percent since 1967. Murder rates in the United States increased from about 50 per million people in 1960 to a high of about 100 per million people in 1980, before gradually declining to rates comparable to those of the early 1960s. But even at our highest rate, back in 1980, the risk of being murdered in the United States was only about 25 percent as high as it was in western Europe in 1450, only 5 percent as high as it is in modern Honduras — the most dangerous country in the world — and only about 2 percent as high as it was in pre-state societies. In today's America, homicides and war deaths together constitute less than 1 percent of all deaths.

The causes of accidental death in the modern United States are also inconsistent with our innate fears. For example, allergic reactions from bees and wasps kill about 50 people per year, dogs and dog bites about 30 people per year, and horse-related accidents about 20 people per year. Snakes, spiders, scorpions, and other insects kill fewer than 50 Americans per year. By comparison, deaths from auto accidents—though at a 60-year low because of seat belts, air bags, and safer cars and roads—are still about 33,000 per year. Obviously, it would be better if American children were innately fearful of automobiles rather than snakes, spiders, and scorpions.

Although many people worry about being killed by a previously unknown assailant as part of a robbery or carjacking, only 20 percent of murders in the United States are committed by people who were previously unknown to the victim, and only 7 percent of all murders are related to attempted robberies, burglaries, or theft. In the United States, about 40 percent of murdered women are killed by someone with whom they'd had an intimate relationship—usually it's a woman of reproductive age who's killed when she tries to end a relationship. In the majority of homicides, men kill other men who share similar characteristics—and may be in the same local peer group—because of a variety of arguments and disputes otherwise unrelated to any violation of the law.

The disagreements that lead to murder may seem rather trivial to others, but they're commonly related to perceived honor, stature, or potential mates. Many of us may recall the television sitcom *Happy Days*—no one messed with Fonzie, who was the undisputed alpha male, and certainly no one would try to make a pass at his girlfriend. As noted by the writer Timothy Beneke, the risk of violence or injury from going after someone else's attractive girl is exemplified by the way we define such a woman—a bombshell, a knockout, striking, drop-dead gorgeous, dressed to kill, or femme fatale.

So the statistics are simple: you're far, far more likely to be killed by someone who knows you, especially someone who knows you rather well and has a serious disagreement with you, than by someone who wants to steal your money or your property. The modern judicial and penal systems, built on about 1,000 years of laws, have altered the dynamics that drove the prehistoric incentives of indiscriminate murder, plunder, and annihilation — at least in most of the industrialized world.

The Current Epidemic of Anxiety and Depression

In any given year, 18 percent of US adults report anxiety, and nearly 30 percent report anxiety during their lifetimes. Nearly a quarter of these cases are self-described as severe, and women are more than 50 percent more likely to report anxiety than men. Anxieties include separation and stranger anxiety, social anxiety, generalized anxiety, panic attacks, and a range of phobias — including fear of heights, open spaces, and closed spaces. Anxieties are commonly diagnosed by self-report or from the testimony of families and friends, because sufferers may not be faced with their specific precipitants when they're in a doctor's office.

According to accepted criteria, about 7 percent of US adults have suffered from a major depressive episode in the past year, and about 17 percent will have such an episode during their lifetimes. And the United States isn't alone. Among seven countries in the Americas and Europe, the estimated prevalence of depression within the past 12 months is consistently above about 4 percent of the population. It may seem hard to believe, but some reports suggest that more than 20 percent of women in the United States between the ages of 40 and 60 have taken an antidepressant.

To meet the criteria for being depressed, people must be in a depressed mood for most of nearly every day or lose interest or pleasure in most of their usual activities for at least two weeks. In addition, they have to have at least three of the following symptoms — significant weight loss or weight gain, not being able to sleep or sleeping too much nearly every day, slowed movement or thinking obvious to others, fatigue or lethargy nearly every day, feelings of worthlessness or exaggerated guilt, indecisiveness or loss of concentration, or recurring thoughts of death or suicide.

Anxieties tend to wax and wane, but they can become sufficiently bothersome and socially debilitating to precipitate sadness and depression. The good news is that most sadness and depression resolve, and they did so even before we had medications. People bounce back — they find new goals and joys, and they reengage — mostly because of support from family and friends.

Unfortunately, however, chronic depression — also called neurotic depression, dysthymia, or melancholy — may ensue if people can't find and achieve new goals. These new goals — whether related to status, honor, financial success, mating, or the like — typically will be different from and less ambitious than the previously unachieved goals that initially precipitated the depression. If people are unable to reorient and refocus, short-term sadness doesn't serve as a useful stimulus to redirect people toward more productive activities but rather becomes a state of chronic low energy and depression. The debilitating helplessness of severe depression makes almost any goal unreachable, and the unremitting misery of persistent depression can ultimately drive its sufferers to suicide.

PTSD: Back to Jason Pemberton

Jason Pemberton was discharged from the army in 2009 after hurting his back in a parachute accident. He and Tiffany moved to Daytona Beach, Florida, where he was studying to be a motorcycle mechanic. Tiffany seemed to be devoted to Jason despite loud arguments frequently heard by neighbors. About six months before the murder-suicide, Jason called the police because Tiffany was causing a disturbance; but he withdrew the complaint when the police arrived.

We don't know what they argued about or whether Jason was worried that his wife would leave him. But we do know that, for some time, Jason had been receiving care for PTSD at a local hospital run by the Department of Veterans Affairs. Tiffany apparently had been told just three days earlier that there was nothing more VA doctors could do for Jason except continue his antidepressant prescription.

PTSD, once called *battle fatigue,* afflicts people who survive and remember particularly stressful situations. The official diagnosis requires more than a month of functional impairment attributable to changes in mood, thoughts, arousal, or reactions in people who have been exposed to clearly defined stress and who reexperience the traumatic event in their minds and alter their behaviors as a result. About 2 percent of all military personnel may experience PTSD, but the rate was up to 10 percent in US combat veterans deployed in Iraq or Afghanistan and is much higher in those who were wounded or saw someone killed. Even about 5 percent of military dogs deployed with American combat troops in Afghanistan developed a canine version of PTSD.

PTSD is exaggerated, memory-based fear that often leads to anxiety and depression. One study of mental health in Nepal showed that the prevalence of anxiety increased nearly twofold after their 2007 war, with PTSD itself found in about 14 percent of the population. This increase in anxiety correlated strongly with an individual's exposure to the impact of the war. Former Nepalese child soldiers were more likely to suffer from PTSD and depression than were those who weren't conscripted to fight.

Although PTSD is a result of key survival traits—memory and fear—it's always been a maladaptation. The self-destructive component of PTSD can lead to suicide, and its overaggressiveness can lead to murder. In the case of Jason Pemberton, we saw both, but sometimes we see the overaggressiveness alone. In My Lai, Vietnam, in March of 1968, American army soldiers killed an estimated 350 or so civilians, mostly women, children, and the elderly. The soldiers, including Lieutenant William Calley, initially may have thought that the village was hiding combatants. But the scale of the persistent massacre dwarfed any imaginable threat, and subsequent witnesses found no evidence of enemy weapons or draft-age men. More recently, in the war in Afghanistan, Sergeant Robert Bales snuck into two villages at night and killed 16 civilians.

In about two-thirds of people with PTSD, symptoms resolve spontaneously and recovery ensues. Others learn to cope over time. But PTSD also can persist—about 10 percent of initially affected Vietnam veterans still suffer from PTSD 40 years later. Exactly why some are more likely to recover than others is unclear, although a supportive social environment certainly helps. But when PTSD persists and is unbearable, stories like Jason Pemberton's become all too common.

Suicide

From an evolutionary perspective, suicide seems counterintuitive. By simple logic, if murder helps perpetuate your genes, suicide would seem to be the antithesis of a survival trait. If we're here because we inherited the behaviors needed to survive, why would humans ever commit suicide?

But suicide probably was always a way for the self-elimination of elderly, disabled, or dependent people who saw themselves as a burden to those who would otherwise have had to care for them. This *altruistic suicide* is exemplified by *sati,* in which an Indian woman commits suicide after her husband's death, and *jigai,* in which the widow of a Japanese samurai warrior would slit her jugular vein with a knife.

For the individual, the pain of submissiveness, sadness, hopelessness, and depression—inadequately treated, untreated, or sometimes only diagnosed in retrospect—is what usually leads to suicide. Common precipitants of suicide include problems with intimate partners, finances, work, or physical health—just the kind of issues that cause depression and are common in people who suffer from PTSD.

Unbearable sadness linked to the fear and loathing of being put into a submissive position isn't the only way to become sad and hopeless enough to commit suicide. But it links directly to the defeat, disappointment, social isolation, and sadness experienced with a loss of pride. Perhaps the best-known example of prideful suicide is hara-kiri, the ritual self-disembowelment performed by defeated Japanese samurai warriors—and captured Japanese soldiers as late as World War II—to avoid humiliation, shame, and possible torture.

Suicides also can be linked to losing social contact with family and friends (*egoistic suicide*), to the failure to achieve desired goals (*anomic suicide*), and to frustration about an inability to fight the system and make things better (*fatalistic suicide*). Egoistic, anomic, and fatalistic suicides all fit with the concept that suicide is driven by feelings of not belonging to a group, loss of key relationships, disappointment in one's own accomplishments, attempts to avoid defeat and disgrace, and the desire to take aggressive control over an otherwise hopeless situation.

Suicide is probably as old as humanity itself, but it's impossible to compare current suicide rates with prehistoric rates. One estimate of Paleolithic suicide comes from the violent hunter-gatherer Hiwis of Venezuela, in whom four documented adult suicides during a seven-year period would be about 2.5 times higher than the modern US rate. The first apparent suicide note dates back to about 4,000 years ago in Egypt. Several biblical figures committed suicide, including King Saul, who killed himself after being injured in battle, Samson, who died voluntarily while saving others, and Judas, who hanged himself in shame. Other famous suicide victims of yore include Antony and Cleopatra.

Suicide is far more common among people who abuse drugs and alcohol. Alcohol and drugs probably directly contribute to the risk of suicide, but some of the association may be because drug and alcohol abuse is precipitated by the same emotions that precipitate suicide.

Suicide is more common in situations in which people think they should be able to compete successfully but fail than in situations in which they never thought success was possible. Consider, for example, what we know about suicide among slaves. Although it's obviously hard to get accurate statistics, available data suggest very high rates of suicide among captured Africans on their way to slavery in America and upon their arrival. Reported suicide rates apparently declined among slaves born into servitude, although they were notoriously high after unsuccessful rebellions. In addition, we don't know how often slaves purposely disobeyed, expecting to be killed — what's now called *suicide by proxy*. A likely interpretation is that soon after their capture, slaves who were confronting the prospect of fear and submissiveness experienced profound anxiety and sadness. By comparison, people born into slavery tended to be more accepting of the submissiveness inherent in the only life they knew — until they raised their hopes, tried to improve their situation, and failed.

For the same reasons, anomic suicide is thought to be more common in affluent societies, where opportunities are theoretically available but where some aspiring achievers get depressed when they don't, for whatever reason, compete successfully for them. In modern societies, economic competition has replaced physical competition as the leading indicator of status. So it's not surprising that suicide rates in the United States rise by about 1 percent for each 1 percent increase in the unemployment rate and peak during economic depression.

Similarly, the suicide rate increased in Russia after the breakup of the former Soviet Union, concurrent with rising unemployment, alcoholism, and murder. Former Soviet states, such as Lithuania, and some strife-ridden African nations, such as Burundi, also have high suicide rates. The affluent country with the highest reported suicide rates is South Korea, where adult rates are in the same range as the Hiwis.

In twenty-first-century America, nearly one million people attempt suicide each year, and someone is successful about every 15 minutes. Although women attempt suicide three times as often as men, men are almost four times more likely to die by suicide than are women.

In the past six years, US hospitals have seen a 50 percent increase in suicide attempts from drug overdose, an even greater increase than predicted by high unemployment rates. If we didn't have such an outstanding emergency medical care system, the successful suicide rate, which has been steady for several decades, would be rising.

Suicide is now the tenth most common cause of death in the United States. As noted earlier in this chapter, it accounts for almost 40,000 fatalities per year, while homicides account for only about 16,000 deaths. The rise in suicide and the fall in murder is evident even in gun-related deaths. The United States has by far the highest gun-related murder rates in the developed world, averaging about 20 times higher than other developed countries. But our 11,000 or so annual gun-related homicides are actually fewer than our 19,000 or so yearly gun-related suicides. In a safer, civilized world, with an expectation that a murderer usually will be brought to justice, violence against others is becoming a less and less attractive course of action. As a result, the anger that used to be directed externally toward other people is now increasingly directed internally, where it's manifested as anxiety and depression—and the resulting aggression is increasingly directed not at others but rather at oneself.

For a prime example of shifting from an environment where you can take out your aggression on others to an environment in which you can't, we need only look to the increasing incidence of suicide among US soldiers. In 2012, about one active-duty US soldier committed suicide per day—accounting for more deaths than those caused by battlefield injuries during the same year. Interestingly, only about 20 percent of these suicide victims were actually in active combat; the others were in situations in which they couldn't legally shoot a person.

Although the trauma of war and PTSD undoubtedly contributes to suicide, studies suggest that more than 50 percent of soldiers who committed suicide had a recently failed intimate relationship, and about 25 percent had been diagnosed with substance abuse. Active American troops used to have lower suicide rates than civilians, presumably because they're prescreened for mental as well as physical fitness. But now, the suicide rate is about 50 percent higher among active military and more than twice as high among military veterans compared with the general population— perhaps because of more drug abuse, disabilities that would have been fatal in prior eras, the nature of modern warfare, and the difficulty of finding a good job back home during a recession. The 13 percent of adult Americans who have served in the military now account for about 20 percent of American suicides.

So it's clear that suicide isn't new and that suicide rates vary with economic conditions and among cultures. When civility and community norms deteriorate, the same stresses that lead to violence can also precipitate suicide—just as they did in Jason Pemberton and in so many other US soldiers.

Modern wars and even genocides, though horrific, still pale by comparison—both in quantity and in barbarism—to the battles of the past. In addition, the social norms we now need in order to thrive in the densely populated, industrialized world have substantially reduced our murder rates. But we're still stuck with a variety of prehistoric survival traits—memories, fears, anxieties, phobias, and an innate tendency to become submissive, sad, and even depressed when confronted with an unwinnable situation.

Now that Americans are more likely to kill themselves than be killed by others—even when considering murders and combat deaths together—the same fears, anxieties, and submissiveness that protected our ancestors in the Paleolithic era are obviously excessive for today's safer world. Modern civilized competitions usually don't revolve around killing or avoiding being killed. We still compete for love, money, resources, and status, but these modern competitions are rarely life-and-death struggles. Nevertheless, losing these competitions can still so adversely affect our ability to thrive and mate that we may become sad, get depressed, and even consider or commit suicide.

For nearly 200,000 years, our ancestors fought a veritable arms race between developing more sophisticated ways to kill and designing better defensive strategies to avoid getting killed. The good news is that the defense is winning. The bad news is that the balance has tipped to the point where our defenses against being murdered are now precipitating the anxiety and depression that are disabling and killing more of us than the violence they were designed to avoid.

Unfortunately, the world remains a dangerous enough place to warrant caution, so it's probably still a bit too soon to turn off our entire repertoire of defensive traits — even if we somehow could do so. That's why anxiety, depression, PTSD, and even the risk of suicide will continue to be the price we pay for our ancestors' success in not getting killed before they passed their genes down to us.

CHAPTER 5

Bleeding, Clotting, and the Modern Epidemics of Heart Disease and Stroke

In August of 2012, the actress Rosie O'Donnell developed an ache in her chest and both arms. Because the 50-year-old O'Donnell had helped lift a heavy woman earlier that day, she thought she must have strained or pulled a muscle. But as the day went on, the pain continued, her skin felt clammy, and she became nauseated and vomited. After she got on her computer and looked up the symptoms of heart attacks in women, she was worried enough to take an aspirin—but not enough to see a doctor until the following day. At that visit, her electrocardiogram showed obvious signs of an acute heart attack. She was rushed to the hospital, where a dye injection showed a 99 percent blockage of the main coronary artery that supplies blood to the heart's left ventricle, which pumps blood out to the body. Doctors quickly inserted a catheter, opened up the artery by inflating a balloon to compress the clot that was blocking it, and inserted a wire mesh stent to keep the artery open and prevent further damage to her heart muscle.

For decades, doctors have euphemistically used the term *widow maker* to underscore the danger of a near-total blockage in this main coronary artery as well as the supposed predilection of men, rather than women, to develop such blockages. But the reality is that coronary artery blockages are the leading cause of death in American women as well as American men, with the only difference being that women tend to get heart attacks an average of about ten years later than men.

Medical science took a great step forward when the British physician William Harvey first described the circulation of blood in 1628. Our circulatory system—which comprises about 60,000 miles of arteries, capillaries, and veins—contains about five quarts (or ten pints) of blood. If this closed circulatory system develops even a small leak, we must immediately plug the dike with a clot to keep from bleeding to death. But if the same clotting system that saves us from such bleeding is turned on when we don't need it, a blood clot—exactly like the one in Rosie O'Donnell's coronary artery—can make us seriously ill or even kill us.

This risk wasn't a big deal in the Paleolithic era, but it is now, because our diets and lack of exercise have made heart attacks and strokes the first and fourth most common causes of death, respectively, in modern America. To appreciate how and why unnecessary clotting has been a key contributor to these modern epidemics, we must begin by understanding how vulnerable we are to bleeding to death and how we're built to prevent it from happening.

THE SURVIVAL CHALLENGE

Our ancestors spent millennia walking or running over the natural terrain of a primitive world. A variety of everyday injuries—slips, falls, cuts, bumps, and bruises—could make them bleed from a cut artery or vein, whether visibly, through the skin, or invisibly, inside their bodies. Even with little or no apparent trauma, a blood vessel can break and bleed—indeed, many of us may notice a black-and-blue mark but not remember the injury that caused it. And on the more extreme end, every time a baby is born, its mother is at risk of bleeding to death when the placenta (sometimes called the afterbirth) tears away from her uterus. If we're all so vulnerable to bleeding, how did our ancestors survive everything from scratches and scrapes to life-threatening punctures, gashes, and childbirth?

At rest, the ten pints of blood in our circulatory system get pumped around our body about once per minute. The good news is that we generally can lose about 15 percent of that blood — about a pint and a half — without any substantial adverse effects. But if we rapidly lose more than that, we can get into deep trouble. If you lose between a pint and a half and three pints within several days, your heart will react by trying to pump the remaining blood around the circuit more rapidly, and your pulse rate will rise. If you bleed more than three pints within several hours, your blood pressure will fall, and many of your organs, especially your brain, may not get enough oxygen to function normally. And if you quickly lose more than about four pints within several hours, you'll need modern emergency treatment to keep from going into shock and possibly even dying.

And it's not only external bleeding that's dangerous. If enough blood leaks out and accumulates internally within the body, a *hematoma* — one of those large black-and-blue bruises we all get — is formed. A hematoma in the thigh can easily contain as much as a pint or more of blood. Hematomas are especially dangerous when they form in parts of the body that can't tolerate their swelling. For example, consider a subdural hematoma — a bruise of the lining of the brain — which can develop after skull trauma. A subdural hematoma puts pressure on the brain by taking up space within the fixed volume of the skull. Just a little more than about three tablespoons of blood in the brain typically causes brain dysfunction, and more than half a cup is often fatal.

To avoid life-threatening and fatal bleeding, our blood needs to clot quickly to plug any leaks. But the clotting system also needs to be perfectly calibrated so that it's activated only exactly when and where it's needed. That's because an unnecessary, misplaced, or larger-than-needed clot can be at least as bad as bleeding itself if it interferes with normal flow through the circulatory system. A clot within an artery deprives downstream organs of needed nutrients and especially of oxygen — and without oxygen, cells in these organs will die. That's exactly what happens when a clot blocks an artery to the heart muscle for more than about 60 minutes and causes a heart attack, or when a clot blocks an artery to the very vulnerable brain for more than about five minutes and causes a stroke. Clots in our veins typically cause less immediate damage than arterial clots but can still have serious consequences.

While blood must flow continuously and smoothly, our clotting systems must stay alert and ready, like a fire station, to respond quickly and decisively only when called upon for a true emergency. It's an incredibly delicate balancing act, and it doesn't take much to throw us off balance.

BACK TO THE PALEOLITHIC ERA

If about 12 percent of our Paleolithic ancestors and about 25 percent of early agriculturalists died from homicide and fatal injuries, just imagine how often they suffered nonfatal cuts, bumps, and bruises. To illustrate the bleeding dangers they faced, let's return to Ötzi the Iceman. Detailed radiological studies of his corpse documented that the arrowhead in his shoulder killed him by severing his left subclavian artery, the major artery that provides blood to the left arm, and causing him to bleed to death. But as if that weren't enough, Ötzi also suffered bleeding in his brain, either from a fall or from a separate blow to the head.

Although injuries were a big concern in prehistoric times, childbirth has probably been an even bigger bleeding challenge in terms of perpetuating the species. A modern American mother who delivers her baby without a cesarean section loses an average of less than a pint of blood when the placenta separates at childbirth, and only about 1 percent of mothers will lose more than two pints if they receive good medical care. But it wasn't always this way. We don't know exactly how much our Paleolithic mothers bled, but we can estimate their risks based on some recent statistics. As late as 1990, it wasn't unusual for at least one woman per hundred to die in childbirth in the poorest parts of the developing world, about one-third of them (or one per 300 deliveries) because of uncontrolled bleeding. If a Paleolithic woman averaged ten deliveries and had the same risks, she had about a one-in-30 lifetime risk of bleeding to death during childbirth or soon thereafter.

Our ancestors needed to be able to clot, and clot quickly, through thousands of generations of childbirth and violence, without modern obstetrical care, blood transfusions, surgeons, or stitches. And today, we're still built to clot the same way.

HOW WE'RE BUILT TO MEET THE CHALLENGE

Blood is pumped out by the heart's left ventricle and distributed to our organs through our arteries. Our smallest arteries then hook up to tiny capillaries, which deliver blood directly to our cells. Blood returns via the capillaries to our veins and then to the right side of the heart. It then gets sent to the lungs to pick up oxygen, returns through the left atrium and into the left ventricle, and then repeats this full cycle, over and over again, about once per minute and about 40 million times by age 75.

Our ten pints of blood include about four pints of red blood cells and six pints of liquid plasma. The red blood cells deliver oxygen, which every cell in our body needs. The liquid plasma carries dissolved salts and nutrients absorbed from the intestine, proteins made by the liver, hormones secreted by various glands and organs, waste products released from the body's cells, metabolic products of cells that have died, and a little bit of dissolved oxygen. Circulating white blood cells — which we need to fight infection — aren't nearly as numerous as red blood cells and don't appreciably contribute to our blood volume.

A red blood cell's function is simple — to pick up oxygen from our lungs and distribute it throughout the body. Each red blood cell contains hemoglobin molecules, which serve as delivery trucks to transport oxygen to our cells. In our lungs, we try to stock the hemoglobin with as much oxygen as possible. On an average pass through our body while we're resting, about 25 percent of the blood's oxygen is delivered to the body's organs and tissues for their immediate use, and the 75 percent that isn't needed returns to the lungs for another go-round. With each cycle, our red blood cells deliver about a gallon's worth of oxygen per minute, or about the amount of oxygen in five gallons of air.

When we exercise, the heart typically can increase its pumping from about five quarts of blood per minute to up to perhaps 20 quarts per minute, partly by increasing the amount of blood pumped with each heartbeat and partly by increasing the number of heartbeats per minute. In addition, our cells can extract about twice as much oxygen from our red blood cells. For most of us, the result is about a fivefold to tenfold increase in the oxygen supplied to the body's cells. But in trained athletes, the heart can actually pump up to six times more blood than it does at rest, and the cells can extract up to three times as much oxygen from each circulating red blood cell — a remarkable eighteenfold increase in the amount of oxygen delivered! Unfortunately, we can't keep that up for long — and if our muscles don't get enough oxygen, they accumulate lactic acid and cramp up. That's why we can run a very long way if we don't run our fastest, but we can't keep sprinting forever.

Our four pints of red blood cells — about 25 trillion in all — account for roughly 25 percent of all the cells in our body. To put this into perspective, each drop of blood has about 250 million red cells. Blood cells are made in our bone marrow—the hollow insides of our bones, especially the long bones of our legs and arms as well as the pelvis, sternum, and ribs. It takes the bone marrow about seven days to make a red blood cell, but at the end of the assembly line, they get churned out at a remarkable rate of more than two million per second.

The normal process for manufacturing red blood cells requires a variety of hormones, vitamins, and minerals. Among the most critical is iron, which is a core component of the red blood cell's hemoglobin molecule. Without enough iron, our bone marrow can't make as many red blood cells, and the ones it does make have less iron and less hemoglobin. So if we're short on iron, it's a double hit to our blood's oxygen-carrying capacity.

A normal red blood cell lives for about 120 days. Abnormal or deformed red blood cells have much shorter life expectancies and, as a result, put more pressure on our bone marrow to rev up production in an attempt to maintain a red blood cell count at or near normal.

If the number of red blood cells declines gradually—because of reduced production, a reduced lifespan, or any cause of slow blood loss—our body compensates by making more plasma in order to maintain overall blood volume. This imbalance results in *anemia,* which is commonly called a low blood count because it's defined by a diminished percentage of red blood cells in a relatively normal overall volume of blood. We can compensate for anemia by increasing our heart's output and increasing the amount of oxygen extracted on each circuit—just as we typically would do with exercise. Each of these two mechanisms can provide a substantial reserve for dealing with a gradual loss of even half of our red blood cells. But if these mechanisms are required at rest, we won't have much, if any, reserve to call upon when we exercise. And that's why fatigue and shortness of breath with activity are classic symptoms in people with marked anemia.

Until your red blood cells diminish by 50 percent or more, it's actually more important to maintain your blood volume than it is to replace the lost red blood cells. That's because your blood pressure — the driving force that circulates the blood — is critically dependent on having an adequate volume in your 60,000 miles of blood vessels. And that's why sudden blood loss is so much more dangerous than a gradual reduction in the number of red blood cells. The rapid decrease in blood volume — both red blood cells and plasma — makes it impossible for the heart to increase its output per minute. Just the opposite happens — with diminished blood volume, the heart's output actually declines no matter how hard it tries to work. And if there's not enough volume, blood pressure falls, the entire system collapses, and we go into shock. So it's critical for doctors to replace the fluid portion of your blood emergently, even if they can't give you a blood transfusion (to bring up the level of your red blood cells) right away.

Smooth Flow

Blood normally flows smoothly through our thousands of miles of blood vessels because the single layer of cells along their inner lining produces and releases a variety of factors that keep blood from clotting, both locally and throughout the circulation. But the cells beneath that single-cell layer don't share the same anticlotting properties — so if the inner lining is damaged and the next layer is exposed, as happens in people who are bleeding (and, as we'll describe later, in people who are having acute heart attacks), the clotting system gets activated.

The rapid-response clotting system includes two largely independent but interrelated pathways. One is based on platelets, which circulate in the blood and rapidly stick like magnets to a specific receptor that becomes exposed wherever the inner layer of cells on the blood vessel wall is damaged. When each platelet binds to the exposed receptor, it releases attractants that rapidly recruit other platelets to join the battle and help plug the dike in the artery or vein.

We normally have about 15 million platelets per drop of blood—at least four times as many platelets as we need to clot adequately after major trauma and at least ten times as many as we need to prevent bleeding after routine damage to our blood vessels during regular daily life. The most common cause of anemia is a low iron level, occasionally because of a poor diet but much more commonly because chronic oozing from the gastrointestinal tract or very heavy menstruation causes us to lose more iron than we can eat and absorb to compensate. When we get iron-deficiency anemia, our platelet counts rise, often to above their usual normal range. What a perfect compensatory reaction—when your blood count is low and you can't afford to lose much more blood, your platelets naturally increase to protect you from more bleeding.

The other clotting pathway depends on more than a dozen blood proteins that participate in a domino-like cascade to form a fibrous mesh. This mesh patches larger tears in a blood vessel's wall and also serves as a matrix onto which platelets can accumulate.

The anticlotting substances that are released by the normal inner lining of the blood vessels destroy or deactivate the very same proteins that help form the fibrous mesh that's so critical for clotting. By doing so, they keep a clot from getting too big and clogging up the entire blood system. And, not surprisingly, anything that reduces the quantity of anticlotting factors or causes them to malfunction can increase the risk of unnecessary clotting—especially in low-pressure, low-flow parts of our veins and especially if we're not very active. Our body's balance between clotting factors and anticlotting factors must quickly shift back and forth to keep blood flowing without clotting unnecessarily or leaking. Any perturbation that disturbs this remarkably delicate balance can cause fatal problems within minutes.

Childbirth

The balance between bleeding and clotting is especially critical during childbirth. A normal placenta weighs a little more than a pound and is attached to the internal lining of the mother's uterus across a length of about ten inches and a width of about seven inches. The placenta has two separate circulatory systems—one from the mother and one from the baby. On the mother's side, the uterine artery creates a network of smaller arteries that nourish the placenta and supply blood to its capillaries. Oxygen and nutrients diffuse across these capillaries from the mother's placental circulation to the baby's placental circulation, without any mixing between the mother's blood and baby's blood. On the baby's side, the umbilical cord connects the baby to the placenta and serves as a conduit that contains the umbilical vein, which brings oxygen and nutrients from the placenta to the baby, and two umbilical arteries, which send blood, pumped by the baby's heart, back to the placenta to be renourished.

The uterine artery carries more than a pint of blood per minute from the mother's circulation to the placenta—a little more than went to Ötzi's arm and about the same amount that flows to an adult's leg at rest. Since we go into shock if we lose more than about three pints of blood within even an hour or so, it's pretty clear why Ötzi's wound was fatal. When the placenta naturally breaks off during childbirth and the uterine artery is severed, why doesn't every woman lose far more than a pint of blood, go into shock, and die within minutes?

First, during pregnancy a mother's red blood cell count increases by about 25 percent, and the volume of her liquid plasma approximately doubles. This extra blood volume not only supports the placenta but also provides a cushion against blood loss at the time of delivery. To make these extra red blood cells that nourish the placenta without themselves developing anemia, pregnant mothers need extra iron and vitamins, which their taste buds don't naturally crave — and that's why multivitamin supplements with iron are now routinely prescribed for pregnant mothers. A potential side effect of the extra salt and water in the liquid plasma is, of course, high blood pressure, as discussed in chapter 3. Pregnant mothers normally counter this risk by becoming naturally less responsive to hormones that would usually raise their blood pressures. As a result, normal blood pressure during pregnancy is actually a bit lower than normal blood pressure at other times. But despite these compensatory hormones, an estimated 3–6 percent of pregnant women develop worrisome high blood pressure. In countries such as those in western sub-Saharan Africa, where current risks may approximate those of the Paleolithic period because prenatal care is minimal and blood pressure medications aren't widely available, high blood pressure causes about half as many maternal deaths as does bleeding.

Second, during the late phases of pregnancy, the mother's liver makes more clotting proteins, thereby reducing the likelihood of severe bleeding when the placenta breaks away. The extra clotting proteins also bring some risks — they increase the *relative likelihood* of unnecessary blood clots by perhaps tenfold compared with nonpregnant women, especially just after delivery, when mothers are less mobile. These clotting proteins and the risk of blood clots don't decline to normal levels for about four to six weeks. The increased probability of unnecessary clotting after delivery (an *absolute risk* of about one per thousand births) and for the prior nine months of pregnancy (about twice that rate) is about the same as the three-per-thousand risk of dying from a post-delivery hemorrhage that existed in many parts of sub-Saharan Africa in 1990. These statistics suggest that the prehistoric clotting system of our ancestral mothers was pretty well balanced for the bleeding and clotting challenges of pregnancy and childbirth, especially at a time when there wasn't any effective way to address either of these challenges.

Third, after the baby is delivered, the same hormones that

stimulated uterine contractions to push the baby out continue to contract the uterus and constrict its blood vessels — both directly and by firming up the surrounding uterine muscle. It's obviously easier to form clots in these smaller, constricted blood vessels.

About 20 percent of the risk of postpartum bleeding is linked to genetic factors. One of the protective factors is a mutation called factor V Leiden, which is the most common inherited muta-tion in clotting proteins. This mutation increases the activity of the clotting protein cascade by a moderate amount in people with one copy of the mutation and by a marked amount in people with two copies. It probably originated in Europe about 30,000 years ago, and it's now found in about 5 percent of people from north-ern Europe but much less commonly in others. Although doctors see the mutation as disadvantageous because of the serious and even fatal clotting it can cause, it also decreases the amount of post-delivery bleeding — a benefit that outweighed those disadvan-tages for our ancestors and explains the mutation's gradual spread back at a time when postpartum hemorrhage was more of a prob-lem than it is today.

Another way to protect against the risk of blood loss and its resulting anemia would be to have a large surplus of iron stored in your body to help you rebuild your red blood cell volume quickly. And a gene mutation called hemochromatosis does just that by increasing the absorption of iron from the food we consume. After first appearing somewhere between 1,500 and 4,000 years ago, probably in Britain or Scandinavia, it was likely spread by the Vikings, especially along the coasts of Europe. People with just one copy of the mutated gene have higher average levels of iron in their blood and in their red blood cells — high enough to help pro-tect them against the iron deficiency that often follows blood loss. And people with two copies absorb even more iron and have even more protection. Given the benefit of extra iron in an era before multivitamin supplements, it's not surprising that almost 10 per-cent of people of northern European ancestry now have one copy of the most common mutation for hemochromatosis—and get at least some of its benefits. About five such people per thousand have a mutation in both copies of the gene and a lot of extra iron. But this double dose is a mixed blessing. People with it—especially men—can gradually accumulate enough extra iron to damage the pancreas, liver, and heart in midlife.

If hemochromatosis provides so much benefit against blood loss and iron deficiency without major side effects until midlife, we might wonder why it didn't spread beyond Europe, where it originated, into Africa. One possibility is that there just wasn't enough interbreeding across the formidable Sahara desert. But we also know that a different hemochromatosis mutation arose independently in Africa, where it never became nearly as common as the European mutations became in Europe. The best explanation is that many infectious organisms—especially the malaria parasite—need to borrow our iron in order to replicate within our bodies. People with low iron levels are relatively protected from these infections, while people with hemochromatosis are more susceptible. So it's easy to understand why extra iron would have a modest net benefit in northern Europe, where the Vikings didn't worry about malaria, but would not have much, if any, benefit in sub-Saharan Africa, where even today malaria kills more than 500,000 people per year and remains the fourth leading specific cause of death, ahead of heart disease, stroke, and cancer.

Our body's historic tilt toward avoiding bleeding to death, whether from the predictable challenge of childbirth or an unpredictable injury, has always been key to our survival. So key, in fact, that even potentially adverse mutations such as hemochromatosis and factor V Leiden spread because their benefits outweighed their side effects. By comparison, no known anticlotting mutation is as common as either of these mutations.

WHEN OUR SURVIVAL TRAITS ARE INADEQUATE

Any deficiency or malfunction of platelets or of any of the more than a dozen clotting proteins can result in abnormal bleeding. But perhaps the best known of these bleeding disorders is hemophilia, most commonly related to a deficiency in a protein called *factor VIII*. The gene for the factor VIII protein is on the X chromosome, and essentially all hemophiliacs are men whose single X chromosome is mutated. Because women have two X chromosomes, they only need one nonmutated X chromosome to provide enough factor VIII for normal blood clotting — and it's exceedingly rare for a woman to have two mutated X chromosomes. As a result, women serve as asymptomatic carriers of the mutation and, on average, give hemophilia and its serious consequences to half their sons (who, historically, have been unlikely to father children) while also passing the asymptomatic mutation down to half their daughters, who then perpetuate the mutation.

The first recorded acknowledgment of abnormal male bleeding was probably in the Talmud, written about 2,000 years ago. By Talmudic tradition, if a baby boy bled to death after circumcision, his subsequent male siblings weren't supposed to be circumcised.

In 1803, John Otto, a Philadelphia physician, described a family in which generations of men but no women suffered severe bleeding. From Philadelphia to as far away as Germany, other physicians soon reported families with serious and even fatal bleeding in males but not females. For reasons that can't logically be explained, the disease was called hemophilia — literally, "love of blood." By the mid-nineteenth century, physicians reported that more than half of afflicted boys died by age seven, and only about 10 percent made it to age 21.

Perhaps the most famous people with hemophilia were the descendants of Queen Victoria of England. Fortunately, only one of her four living sons developed hemophilia, and only two of her five living daughters passed it on to their sons. But her female offspring married royals throughout Europe, ultimately passing the disease to noble families in Prussia, Spain, and Russia, including Crown Prince Alexei, the son of Tsar Nicholas II, whose entire family was killed in the 1918 Russian Civil War. Queen Victoria's family tree exemplifies how asymptomatic mothers and sisters can perpetuate an X-chromosome mutation that's fatal in men.

Even in the absence of inherited disorders of clotting proteins or platelets, some of us can also bleed from protein or platelet deficiencies that we acquire during our lifetimes. For example, antibodies may inadvertently destroy a clotting protein or our platelets—by using the same mechanisms that they're supposed to use to attack viruses, parasites, and bacteria. In others, damage to the liver (as seen in people with cirrhosis resulting from chronic viral hepatitis or alcohol-induced liver failure) can reduce the production of clotting proteins. Similarly, damage to the bone marrow (as seen in patients who receive toxic chemotherapy to treat leukemia or other cancers) can reduce its production of platelets. Although we normally have more than enough clotting proteins and platelets circulating around, ready to jump into action when needed, sufficiently severe reductions can prohibit us from clotting normally and cause us to bleed after even the most minor injuries.

Too Much Clotting

Just as too few clotting proteins or platelets can lead to excessive bleeding, too many clotting proteins or platelets can lead to excessive clotting. These clotting tendencies, often inherited but also commonly found in pregnant women and in some people with cancer, especially predispose us to clots in our veins.

Blood flows more slowly in our veins than in our arteries, and any vein has the potential to clot and cause problems. For example, blockage of the veins that drain the kidney raises pressure in its capillaries, and the resulting damage can cause severe kidney dysfunction. Clots in key veins draining blood back from the brain to the heart can cause severe headaches, strokelike symptoms, and seizures.

But by far the most common site for vein clots is in our legs. The reason is pretty simple—in our leg veins, blood has to flow uphill, against gravity, back to the heart whenever we're standing or sitting, and it also needs to deal with kinks in the veins when we're sitting. For example, about 3 percent of us will develop detectable blood clots in our legs after an eight-hour plane trip, and the risk goes up if the flight is even longer. Muscle activity in our legs serves as a bit of a pump, even though the resulting pumping action is only about 5 percent as strong as the forward pumping by the heart muscle. Fortunately, that 5 percent is enough to keep blood flowing smoothly if our leg muscles remain minimally active—for instance, when we flex our calf muscles or get up and walk around even briefly.

When clots form in superficial leg veins that we can see, we call them varicose veins. Fortunately, varicose veins usually cause nothing more than cosmetic embarrassment and mild discomfort. By comparison, clots in the larger, deep veins, developing either on their own or by extension from a superficial vein, can get a lot bigger and can cause substantial discomfort and even swelling of the leg. But what's especially dangerous is if a piece of one of those larger clots, especially a deep clot that began or has extended into the thigh, breaks off and travels through increasingly larger veins back to the right side of the heart. Typically the clot will pass easily through the right atrium and right ventricle into the pulmonary artery. However, as the branches of the pulmonary artery become progressively smaller, the clot will lodge and obstruct blood flow to the lung. This serious and sometimes even fatal complication is called a pulmonary embolism.

Although clots in our veins can be dangerous, clots in our arteries, which take blood from the heart to all our organs—

much more rapidly and under much higher pressure—are even worse. In our arteries, even an otherwise perfectly normal clotting system can get activated despite perfectly normal flow—not because of a true laceration of a blood vessel but rather because of damage to the single layer of cells in its protective inner lining. And by far the most important underlying explanation for such damage is atherosclerosis.

Atherosclerosis is the deposition of cholesterol-rich fat that adheres to the inner lining of our arteries. The average American probably eats about 200–300 milligrams (about 1/100th of an ounce) of cholesterol per day, but our liver makes more—three to five times as much as we eat. Cholesterol is critical to our health—it's part of the membrane of each of our 100 trillion cells, it's one of the building blocks of many of our body's hormones, and it's also part of our bile, which we need to absorb fats as well as vitamins A, D, E, and K.

Interestingly, the amount of pure cholesterol we eat has very little impact on our blood cholesterol levels. As noted in chapter 2, our hunter-gatherer ancestors' animal intake probably provided at least as much cholesterol as a modern diet. What's different is that our ancestors ate wild animals, whose lean meat is less fatty and whose fat is higher in healthier polyunsaturated fats. By comparison, modern diets are rich in the naturally saturated fats of domesticated animals and may contain the artificially created trans fats of hydrogenated vegetable oils. And it's these fats, not the actual amount of cholesterol we eat, that cause the problem.

It goes like this. The fats we eat are absorbed by the small intestine. These fats—called lipids—don't dissolve directly in the liquid plasma because they're not water soluble. They need to be carried around by water-soluble *lipoproteins*—shipping containers that are made in our livers. Some fats are shipped directly from our intestines to our organs within minutes to provide immediate fuel, and some go right to our fat cells to be stockpiled in case of future starvation. But most fats are carried around by lipoproteins in our blood plasma to provide a steady source of fuel to our organs in between meals and throughout the day and night.

These circulating lipoproteins initially contain a lot of fat. They carry *triglycerides*—which provide most of the fuel—and cholesterol, some of which is transported to cells that need it and some of which is part of the lipoprotein shipping container. The lipoproteins are attracted to various parts of the body by specific receptors that act as magnets, telling them where they should temporarily dock to release their cargo of fuel-rich triglycerides as well as whatever cholesterol may be needed. Little by little, the lipoproteins are depleted of much of their cargo—most of their triglycerides and some of their cholesterol. But they retain their shipping containers, which are composed mostly of protein and cholesterol. As a result, they become low-density (low-in-triglycerides) lipoproteins, or *LDL*. The cholesterol carried by LDLs (LDL cholesterol) is termed *bad cholesterol* because excesses that aren't removed by the liver tend to get deposited in the walls of arteries. As the fatty deposits accumulate, they form progressively larger streaks and plaques.

The amount of deposited cholesterol depends on how much saturated fat we eat, since this is the fat that requires the liver to build transporting lipoproteins. But our cholesterol levels are even more dependent on the liver, which senses the level of fat in the bloodstream. Then, like a thermostat that needs to know when to turn on the furnace, the liver manufactures the amounts of cholesterol and lipoproteins it thinks we need.

This basic metabolism explains why bad cholesterol levels are often elevated in Western societies: we eat too much saturated fat and stimulate our livers to produce more cholesterol-rich lipoproteins to carry them. To make matters worse, our livers often make more cholesterol and lipoproteins than we require because, in most of us, the liver doesn't turn off production as well as we might like. This tendency to overproduce lipoproteins was probably important in the Paleolithic era. Back then, our ancestors needed plenty of lipoprotein shipping containers to carry all the fats they occasionally absorbed after gorging on a huge meal and to keep those fats circulating in the blood as ready sources of fuel when they couldn't get enough to eat. Now, however, this misset thermostat leads us to make far more lipoproteins and bad cholesterol than we need.

We can reduce bad cholesterol levels by eating less saturated fat or by fooling our livers into making less cholesterol and lipoprotein. Statins, the most common medication used to treat elevated cholesterol levels, inhibit the liver's ability to manufacture cholesterol and, secondarily, increase the activity of the specific docking receptors that remove LDL cholesterol from the blood. Newer medications reduce our ability to absorb cholesterol in our intestines or increase the number of cholesterol-removing receptors in our livers.

High-density lipoprotein, or *HDL,* is composed mostly of protein and has little in the way of fat. It acts as a Dustbuster — picking up extra cholesterol from tissues and blood, then taking the cholesterol back to the liver, where it can be made into bile or recirculated later as needed. When you get a blood test to measure HDL, it isn't really measuring the HDL itself but rather the amount of cholesterol being carried by it, commonly called *good cholesterol.* Your total cholesterol level includes LDL cholesterol, HDL cholesterol, and cholesterol that comprises or is being carried by other lipoproteins.

When cholesterol and other lipids accumulate on the internal lining of our arteries, and especially our coronary arteries, these fatty deposits narrow an artery but rarely totally block it all by themselves. Instead, they form surface plaques that are predisposed to fissure and rupture, which then causes a tear in the single-cell lining of the artery and exposes the underlying arterial tissue. When this happens, the artery does what it's supposed to do — sends out all the signals that bring platelets and clotting proteins to repair the rip.

The subsequent clot not only plugs the leak, it can also partially or even totally block blood from flowing farther down through the artery. So even though we're not going to bleed to death from a fissure in the one-cell-deep inner lining of an artery, we'll be in big trouble if a clot blocks so much of the downstream blood flow that cells die in the organ whose blood supply is compromised. And even if an arterial clot doesn't completely block blood flow (in that case, it would be called a nonoccluding clot), it can still cause problems. First, a nonoccluding clot may block enough blood to deprive downstream tissues of the oxygen they need, especially during exercise. Second, if some of the clot breaks off and travels downstream, it can completely block a smaller artery.

As discussed in chapter 3, when atherosclerosis and clots compromise arterial blood flow to the kidneys, they cause high blood pressure. When they block flow to our legs, we get cramps when walking. And when atherosclerosis and clots get very advanced, they can block blood flow to our intestines and cause cramps when we eat. But the two most serious problems arise when atherosclerosis and arterial clots cause heart attacks and strokes, two conditions that were rare in the Paleolithic era but now plague the industrialized world.

THE MARCH OF CIVILIZATION

Starting about 6,000 years and about 300 generations ago, the optimal tuning of our clotting system began to change slightly when our ancestors developed new techniques to avoid bleeding to death. Sutures — stitches, in lay terms — may first have been used in Mesopotamia, where a number of surgical procedures were initially described. At least 3,500 years ago, ancient Egyptians treated wounds with a combination of lint to promote closure, animal grease to provide a barrier, and honey as an antiseptic agent. The Romans were the first known users of tourniquets to control severe bleeding. And in 1920, the first adhesive bandage — patented as the Band-Aid — was invented by Earle Dickson, an employee of Johnson & Johnson.

But for rapid or severe blood loss, the first goal is to restore at least some of the blood's volume, and the second goal is to replace the red blood cells themselves. The first influential attempt to restore blood volume quickly, even without transfusing red blood cells, took place in the 1830s, when Thomas Latta of Scotland reported that infusion of an intravenous salt solution could save one-third of his cholera patients. In 1876, Sydney Ringer of London developed his now-eponymous solution, in which salts are balanced to mimic their concentrations in blood. By 1931, safe intravenous salt-and-sugar solutions were manufactured and widely available when needed.

The first recorded successful blood transfusion was performed in dogs by English physician Richard Lower in 1665. Two years later, he and at least one other physician transfused a small amount of sheep's blood into a human without any obvious benefits or apparent adverse effects. But subsequent attempts led to serious complications, so the field lay dormant until 1818, when British obstetrician James Blundell successfully transfused four ounces of blood from a husband into his wife, who was suffering from a postpartum hemorrhage. Unfortunately, further attempts in other patients brought Blundell only one success and many failures.

We now know why these early attempts at blood transfusions often failed: transfusions cause severe or even fatal allergic reactions if the donor's blood isn't compatible with the recipient's blood. Blood transfusions really weren't practical until the early 1900s, when blood typing was first understood, after which things progressed very quickly. By 1914, blood could be preserved for later use for a compatible recipient. The first hospital blood bank was established in 1937, and by World War II the Red Cross had developed a storage and distribution system to make blood routinely available for lifesaving transfusions.

Despite all the attention paid to bleeding over many millennia, little if any attention was given to the other side of the coin — the downsides of too much clotting. Leg swelling and varicose veins had been noted at least since Greek and Roman times, but no one knew what caused them.

The first clinical recognition of swelling directly related to a blood clot in a deep leg vein probably occurred in 1271. More cases were reported over the next several centuries, but doctors still had no idea what was going on until the 1600s, when William Harvey described how blood circulates through our arteries and veins. By the late 1700s, the concept of venous blood clots, which could enlarge and expand, was understood. The first description of an apparently fatal pulmonary embolism wasn't published until 1819, and accurate antemortem diagnosis of pulmonary embolism wasn't possible before the 1960s.

Heart Attacks and Strokes

On the arterial side, William Heberden first described coronary artery disease, which is the underlying cause of nearly all modern heart attacks, in 1768. His patient had what he called *angina pectoris* — a cramp in the chest caused by inadequate blood supply to the heart's muscle during exercise. Within the next 25 years or so, other British physicians reported that patients with angina often died suddenly, that their autopsies showed what were described as cartilaginous narrowings in their coronary arteries, and that at least one patient had indisputable autopsy evidence of damaged heart muscle, indicative of a heart attack. At about the same time, the anatomist Giovanni Morgagni in Italy described an obese woman who died suddenly and had autopsy evidence of whitish arteries and a ruptured heart, undoubtedly from an acute heart attack. But it wasn't until 1878 that a doctor diagnosed a heart attack in a living patient and confirmed it subsequently by autopsy.

Heart attacks were so unusual that one London hospital found evidence of a recent heart attack in fewer than two autopsies per year in the early 1900s, compared with nearly seven times as many in the late 1940s. In the early twentieth century, the antemortem diagnosis of a heart attack was so rare that Paul Dudley White—the US cardiologist who later cared for President Eisenhower during his heart attack—stated that he had never even heard about heart attacks when he graduated from medical school in 1911.

In retrospect, it's surprising that coronary disease, though unusual, wasn't clinically recognized sooner. Arterial calcifications, which can be seen by x-ray in the plaques of chronic atherosclerosis, were described in Italy in 1575—a full 200 years before Heberden. Using x-ray technology, we now know that about one-third of mummies from ancient Egypt and pre-Columbian Peru had calcifications in their arteries. Even Ötzi the Iceman had calcification in his arteries and genetic evidence of an increased risk of coronary artery disease.

The concept that a heart attack was caused by an acute clot superimposed on atherosclerosis and not by the atherosclerosis itself was first recognized by the physician James Herrick in 1912. Since not all autopsies of fatal heart attack victims showed clots, it took a while before physicians figured out that the offending clots often were subsequently partly or totally dissolved by the body's anticlotting proteins—although not quickly enough to save downstream heart muscle from dying. We now know that nearly all heart attacks are caused by clots and that nearly all those clots are triggered by fissure or rupture of atherosclerotic plaques in the coronary arteries.

As the twentieth century progressed, advanced atherosclerosis became increasingly common. By the Korean War, more than three-quarters of young soldiers killed in combat already had atherosclerotic plaques in their arteries, and such plaques were nearly ubiquitous in the arteries of middle-aged Americans.

What we now call a stroke was called *apoplexy*—literally, Greek for "striking down"—by the ancient Greeks and Romans. The famous Greek physician Hippocrates attributed apoplexy to stagnation of the blood long before Harvey's theory of circulation. By historical accounts, Roman emperor Trajan reportedly suffered from sudden paralysis—consistent with a stroke—soon before he died in 117 CE, and Pope-elect Stephen of Rome died from apoplexy in 752.

By the mid-1600s, English physicians attributed apoplexy primarily to acute bleeding in the brain. But by two centuries later, physicians realized that apoplexy could be caused by clots that blocked blood flow to the brain as well as by bleeding into the brain.

THEN AND NOW

Our finely tuned system, with its delicate balance between smooth flow under normal conditions and rapid clotting when necessary, developed over thousands of years—during which our ancestors were extremely active. They rarely lived long enough to develop clots in their pulsating arteries, and they were active enough to avoid most blood clots in their veins.

Just as our Paleolithic clotting system couldn't always save our ancestors, it still isn't always good enough to save us from really major injuries. Sometimes an injury is simply too severe—for example, when there's a massive cut in a vein or artery that mere clotting can't fix. Injuries remain the fifth leading cause of death in sub-Saharan Africa, where about four significant injuries occur per every ten people each year, and about 1 percent of those injuries are fatal. Unintentional injuries still account for about 7 percent of all worldwide deaths, and the rate is even higher in many developing countries.

And bleeding is also still a challenge in modern America. Consider president Ronald Reagan, who nearly bled to death after being shot in 1981, despite his normal clotting system. Reagan quickly went into shock when he bled well more than half his blood volume from the bullet wound near his main pulmonary artery. He was saved by the emergency infusion of an intravenous salt solution to restore his blood volume, the transfusion of eight units of red blood cells — which is the equivalent of the number of red blood cells in eight pints of blood — to help carry oxygen to his organs, and surgery to sew up the leaking artery.

Interestingly, Reagan's injury and survival are part of the story behind the declining murder rates in the United States over the past 30 years. One reason why murder rates have declined is better medical care — ranging from better ambulances to well-equipped modern trauma units in hospital emergency rooms. As a result, people who suffer from a gunshot wound, a knife wound, or an injury from another weapon are now less than half as likely to die as they were several decades ago.

Although the risk of postpartum hemorrhage is now less than 2 percent of what it was in many developing countries just 20 years ago, that doesn't mean the risk is too low to be ignored. Modern medical management of the final stages of labor includes administering a drug to increase the tone of the uterus's muscle just about when the baby's shoulders are delivered (so the uterus can shrink faster after delivery and squeeze its blood vessels to keep them from bleeding as much), gentle traction on the umbilical cord (to aid the separation of the placenta from the uterus), clamping and then cutting the cord after about 60 seconds, and massaging the uterus after delivering the placenta. The combination of these treatments reduces the risk of postpartum hemorrhage by about 70 percent, with about half of that benefit coming from the medication given to increase uterine tone. The World Health Organization now recommends that all women be given either intravenous or intramuscular medication — such as oxytocin (brand name

Pitocin)—to help the uterus increase its tone and clamp down bleeding vessels. Additional oxytocin, sometimes supplemented by other medications, is recommended if peripartum bleeding looks like it might become excessive.

Despite all these measures, about three out of 100 women in America still suffer from postpartum hemorrhage. Fortunately, because of their extra red blood cells and the intravenous fluids they routinely receive in the hospital, blood transfusions aren't generally needed unless the mother has lost four or more pints of blood. As a result, only about 10 percent of women with postpartum hemorrhage require blood transfusions. Now, about 20 American mothers per million die from postpartum hemorrhage, a rate that's less than 1/150th the risk in the Paleolithic era.

Because of sutures, blood transfusions, less violence, and better obstetrical care, we're now at far lower risk of bleeding to death than ever before in human history. However, we're the descendants of people who were really good at clotting, and the genetic tendencies we inherited from them are directly implicated in the most common causes of death in the industrialized world. In modern America, diseases caused by clots—heart attacks, clotting strokes, pulmonary embolisms, and the like—cause about 25 percent of all deaths, more than four times the number of deaths caused by all forms of bleeding, including trauma, murder, suicide, bleeding strokes, ulcers, and other bleeding conditions. And the genes that cause this imbalance will continue to be perpetuated because these clotting diseases rarely strike before we've successfully had children.

Heart Attacks: Back to Rosie O'Donnell

If you have symptoms potentially consistent with a heart attack, it's critical to get to an emergency room as quickly as possible. It's far better to call 911 and be taken by an ambulance that can provide needed care than to wait around for someone to drive you there. In the emergency setting, the most important diagnostic test is the electrocardiogram, which will be performed immediately, even as nurses and doctors check your pulse and blood pressure.

If the electrocardiogram shows evidence of a heart attack, you'll usually be whisked off to a catheterization laboratory to have dye injected into your coronary arteries and see if one of them is, in fact, blocked. If so, you'll be treated like Rosie O'Donnell. If you're in a hospital without a catheterization laboratory, you'll either be given a clot-busting drug or be transferred to one that has such a laboratory. If doctors aren't sure about the diagnosis, they'll usually watch you for six hours or so, repeating the electrocardiogram as well as blood tests to see if they show any evidence of damage to your heart muscle.

Legend has it that heart attacks in women produce importantly different symptoms from those they produce in men. But the reality is that all people with heart attacks may get chest discomfort that radiates to their arms, neck, or jaw; shortness of breath; sweating; weakness; dizziness; and nausea. People who are inactive are less likely to have warning signs—such as the classic effort-induced angina described by Heberden—than active people. That's because angina is precipitated by inadequate blood supply to the heart muscle during exercise, despite an adequate supply at rest. Not surprisingly, this symptom of partial but not complete blockage of a coronary artery is less common in people who are less active. Since women, on average, get heart attacks about ten years later than men, they're oftentimes less active and, hence, less likely to have prior angina symptoms. And heart attacks in older patients, both women and men, are more likely to be heralded by shortness of breath, confusion, or other nonspecific symptoms than are heart attacks in younger patients.

In my own studies of more than 18,000 people coming to emergency rooms, symptoms in women really aren't all that different from symptoms in men. The biggest difference is that when a man gets any of the symptoms mentioned above, everyone's first concern is a heart attack. When a woman gets these symptoms, people are less likely to think about a heart attack and more likely to try to blame it on something else — muscle cramps, heartburn, stress, or whatever.

If doctors are sufficiently sure that everything's okay — whether they know right away or after about six hours of observation — you'll be sent home, usually with a recommendation to have further testing, such as a stress test, either immediately or within a week or so. If the evaluation shows other evidence of coronary artery disease, you'll need a more detailed evaluation. Regardless, many people find the entire experience to be the kind of wake-up call that helps them address any reversible risk factors they may have.

The main risk factors for heart attacks are elevated levels of the bad LDL cholesterol, low levels of the good HDL cholesterol, smoking, diabetes, high blood pressure, older age, and being a man. Heart attacks are also more likely to occur in people who have evidence of inflammation, perhaps because the inflammatory process makes a cholesterol-laden plaque more likely to fissure.

We can be sure that Rosie O'Donnell's treatment included recommendations to lose weight and exercise more — especially because these lifestyle changes will lower bad LDL cholesterol levels, raise good HDL cholesterol levels, and reduce the severity of obesity-related diabetes. Heart attack survivors are also routinely prescribed a statin to reduce LDL cholesterol down to levels similar to those of modern-day hunter-gatherers — an LDL cholesterol level below 70 milligrams per deciliter (mg/dL).

Reducing cholesterol levels can stabilize existing plaques, reduce their further growth, and sometimes even shrink them. When plaques are more stable, they're less likely to fissure, rupture, or precipitate abnormal clotting in our arteries. But no lifestyle alteration can change our blood's intrinsic ability to clot. That's why the simple aspirin that Rosie O'Donnell took before seeing a cardiologist may well have saved her life by preventing the clot from totally obstructing her coronary artery. To keep her stent open, though, she would need more powerful medications to inhibit platelet function. Blood thinners that inhibit platelets or clotting proteins have become key components of the comprehensive treatment of people who have had or are at high risk of having abnormal clotting; they slightly increase the risk of bleeding while markedly reducing the risk of life-threatening clots.

Strokes

As best as we can tell, more than three-quarters of strokes occurring between 1877 and 1900 were caused by bleeding. But in the twenty-first century, about two-thirds of strokes in Japan and more than 80 percent of strokes in the United States and western Europe are caused by clots. The reasons are twofold. First, by far the leading cause of bleeding strokes in prior eras was high blood pressure — remember Franklin Roosevelt? — which now can be treated and controlled. Second, the rise in atherosclerosis has led to more plaques and clots in the arteries that supply blood to the brain.

The warning symptom for potential stroke is called a *transient ischemic attack* — or a mini stroke. Brain function — usually a specific function such as speech or the ability to move a part of the body — is temporarily lost, but the patient makes what appears to be a complete recovery. In a full-fledged stroke, the deficit persists for more than a day and perhaps forever. Because modern medical care can reopen a blocked brain artery or repair a bleeding blood vessel in the brain, patients with a suspected stroke should follow the same 911 protocol recommended for suspected heart attack victims.

When a clot blocks an artery to the brain and causes a stroke, the damage can be attenuated if doctors can rapidly restore enough blood flow to keep the downstream tissue alive. That's why doctors administer clot-dissolving drugs—if possible, within 90 minutes of the first onset of symptoms. No treatment can save brain tissue that's been totally deprived of blood for even five minutes, but surrounding brain tissue that's receiving some but not sufficient blood flow can hang in there for a while and can be saved if enough blood flow is restored within three to four hours or so.

Despite the many similarities between heart attacks and strokes, there are several key differences. Strokes are unlikely to be caused by the complete obstruction of a large carotid artery—which supplies blood to the brain—at the site of a plaque fissure, presumably because this artery is substantially wider than the coronary arteries. Instead, strokes are more likely to be caused by a piece of clot or plaque that breaks off from that fissure site and moves until it blocks a smaller artery within the brain itself. To prevent a stroke in patients who have had a stroke or mini stroke—or even in people found to have narrowing of the carotid arteries—antiplatelet agents as well as agents that block clotting proteins are useful. Patients at risk for stroke should also receive medication to normalize their blood pressure and a statin to lower their cholesterol levels. In addition, patients who have a symptomatic, severe partial blockage of a carotid artery (as in a transient ischemic attack) will often benefit from an endarterectomy, a procedure in which a surgeon literally peels off the atherosclerotic plaque that lines the inside of the artery.

A second key difference between heart attacks and strokes is in their emergency treatment. For a heart attack, a three-step sequence—angioplasty (blowing up a balloon to compress the clot and open the artery), the insertion of a metal mesh stent to keep the artery open, and a cocktail of drugs to keep the clots from returning—has proven to be both more efficacious and less likely to cause bleeding elsewhere than relying on the most powerful clot-busting drugs alone. In the arteries to the brain, however, angioplasty and stenting have not been as successful—probably because of the hazards of having a small piece of a clot break off and cause even more problems farther downstream. Newer, more effective technologies that emergently aspirate the clot and reopen the artery are just becoming available, but the usual treatment for acute stroke at most hospitals is still clot-busting drugs, not catheter-based interventions.

A third difference lies in how well the brain and heart can compensate for the damage caused by arterial blockage. Our hearts have rather large reserves, so they can still pump enough blood to keep us alive and kicking even if 20–30 percent of the heart muscle dies. In the brain, however, each specific region performs a unique function that typically isn't duplicated by another region. If the region of the brain that moves your right side dies, you'll be paralyzed, and even aggressive rehabilitation probably will be able to restore only some of that function.

Vein Clots and Pulmonary Embolism

The main challenge for our veins is that we were built for constant motion, not for our relative physical inactivity in the modern world. Consider the case of a young British video-game fanatic named Chris Staniforth, who spent 12 or more hours each day sitting at his computer. He died in 2010 when blood clots formed in his legs, broke loose, and became a fatal pulmonary embolism. His platelets and blood proteins, which had been fine-tuned over millennia to avoid improper bleeding or clotting in a continuously moving stream of blood, formed spontaneous clots in a stagnant pool of blood in his legs. That's why you should periodically stretch and walk when you're taking a long plane, car, bus, or train ride—to stimulate blood circulation and avoid spontaneous clotting. For long plane rides, compression stockings may provide additional protection. Unfortunately—or perhaps fortunately for your pocketbook—your veins won't be any happier in business class than in economy. We don't know whether fully reclining sleeper seats may be better, since the blood in our veins doesn't have to fight gravity when we're horizontal. But even then, it's important to get up and move around rather than just alternate among various degrees of sitting and lying down.

If you think you may have a clot in a leg vein, you should see a doctor quickly — it's not a 911 emergency, but you shouldn't fool around. Symptoms include pain and swelling in the calf or sometimes even the thigh — usually obvious because the other leg typically isn't involved. The doctor will check you out, generally with an ultrasound examination and often with blood tests. If you have evidence of a clot, you'll receive medication to tamp down your clotting proteins so your normal anticlotting proteins can dissolve the clot over several days.

The suspicion of a pulmonary embolism, on the other hand, requires urgent evaluation, typically with an x-ray scan of the lung's arteries. Depending on the size of the affected lung segment and the amount of damage it suffers, symptoms can range from none to shortness of breath to coughing up blood to life-threatening or even fatal reductions in the ability of the lungs to oxygenate your blood. To minimize the damage, patients are routinely given medications to reduce clotting proteins, may also be given clot-busting medications, and, on rare occasions, can benefit from surgery to remove a life-threatening clot.

Back in the Paleolithic era, when the human clotting thermostat was being set, it wasn't important to worry about what would happen millennia later, when long car rides and airplane flights would cause blood clots to develop in the veins of our legs with the potential to move to our lungs, where they could be fatal. Nor could it be anticipated that changes in diet, activity, and blood pressure would narrow our arteries and predispose us to form clots that could precipitate heart attacks and strokes.

We still need to clot—not only to survive serious injuries but also to deal with childbirth and the wear and tear of everyday life. But our natural delicate balance between bleeding and clotting now clearly tilts toward too much clotting. That's why excessive clotting causes more than four times as many deaths in modern America as can possibly be attributed to bleeding and why heart attacks and strokes are now two of our four leading causes of death.

To understand just how much things have changed since the Paleolithic era, consider what's happened to the risks associated with pregnancy and childbirth. In modern America, it's clotting in the veins and arteries—not bleeding—that's the leading cause of maternal death. And despite the ready availability of modern medications to treat it, high blood pressure kills more pregnant women in high-income parts of North America than does postpartum hemorrhage. What a remarkable turn of events!

PART II

PROTECTING OUR BODIES IN THE MODERN WORLD

CHAPTER 6

Can Our Genes Evolve Fast Enough to Solve Our Problems?

Since 1990, the life expectancy of non-Hispanic white Americans without a high school diploma has declined by four years, with a greater decline among white women than white men. Although these declines are not as large as those seen among men in post–Soviet Union Russia or in parts of Africa that have been devastated by the HIV/AIDS epidemic, they are unprecedented in the United States. Since the beginning of the Industrial Revolution, we've taken for granted that human life expectancy, with rare and isolated exceptions, is progressively lengthening, generation after generation.

This striking decline in life expectancy in modern America raises an obvious but critical question. Is it just a temporary or isolated anomaly, or will improvements in living standards paradoxically decrease life expectancy as obesity, diabetes, high blood pressure, depression, anxiety, heart attacks, and strokes all increase?

To get a sense of the future, it's important to first go backwards and understand human reproductive success and life expectancy during and after the Paleolithic era. Then we'll look at current data to help us predict the future.

BACK TO THE PALEOLITHIC ERA

The human species is perpetuated by those who are most successful at reproducing and maximizing the number of their offspring who live to do the same—not by people who live the longest or even necessarily stay the healthiest. Fitness and longevity matter only if they help us attract more or better mates, increase fertility, or protect our offspring from predators and hazards.

To self-perpetuate, a species needs to have survival rates that outstrip its death rates. To achieve this goal, the three requirements are: a high enough annual birth rate, enough children living into adulthood and having children of their own, and enough of those children living into adulthood to repeat the cycle. Scientists have calculated that self-perpetuation requires the annual adult survival rate to be at least 70 percent as high as the effective birth interval (the average interval between the births of children who survive to reproductive age). In other words, let's say a baby is born every two years, and half the children survive through adolescence (an effective birth interval of four years). If adults in their childbearing years have less than an 8 percent annual mortality rate (which is arithmetically equivalent to a survival rate of slightly above 70 percent after four years), the species can keep going.

Now let's look at what we know about each of these considerations among our ancestors—fertility, survival through adolescence, and life expectancy. We don't have any information about fertility rates in the Paleolithic era, but we can get some insights from modern hunter-gatherer societies. For example, Gambian village women and the Aché women of Paraguay have their first babies at around the age of 18, an age that seems to provide the best trade-off between their own full growth and fitness on the one hand and the number of their reproductive years on the other hand. Gambian women and isolated Dogon women in Mali typically have between seven and 13 live births, with about half the babies surviving to age ten. More pregnancies, or having too many children too quickly, would result in diminishing returns because of the difficulty in feeding, rearing, and protecting them. Women control their reproduction in a variety of ways, including deciding how long they will breastfeed their babies (fertility is markedly suppressed by nursing). All these behaviors — when these women first get pregnant, how many children they have, and how they space those children — are remarkably consistent with statistical estimates of how they should maximize their net reproductive success and pass on their genes to as many surviving children and subsequent descendants as possible.

In terms of survival through childhood, extensive analyses of ancient human skeletal remains show that about 50 percent of our Paleolithic ancestors died before the age of ten. But once they made it to age 20, they had about a 35 percent chance of making it to age 40. If we look at isolated hunter-gatherers observed during the twentieth century, we see similar but slightly better statistics: about 45 percent die by age 15, but of the 55 percent who make it to age 15, about two-thirds (or 35 percent overall) make it to age 45 — old enough to be grandparents. And those 35 percent who make it to age 45 survive an average of about another 20 years. If you do all the arithmetic, average life expectancy at birth is about 30 to 35 years.

Based on these calculations, Paleolithic fertility rates were high enough and mortality rates during peak childbearing years low enough to perpetuate *Homo sapiens,* despite a 60 percent or so childhood mortality rate. By comparison, studies of skeletal remains of Neanderthals show not only a somewhat higher rate of death before age ten but also as high as an 80 percent chance of death between ages 20 and 40 — an annual mortality rate of more than 7 percent, just at the margin of what would have been required to sustain their species. These data explain why Neanderthal populations were always relatively small and near the brink of extinction, while our *Homo sapiens* ancestors were able to grow their population during good times and maintain a cushion to avoid extinction during challenging times.

We're also helped by our longevity beyond the usual childbearing years. Although natural selection isn't driven by what happens to people after they have completed procreation and childrearing, healthy elders are good to have. Middle-aged and older women have always contributed substantially as grandmotherly helpers in childrearing. And although men reach their peak physical strength by their mid-twenties, hunter-gatherer men don't reach their peak hunting skills — and hence maximize the net number of calories they can generate beyond their own needs to help feed their families — until about age 40 or 45. In addition, men can continue to father children into old age.

THE MARCH OF CIVILIZATION

Human fertility, childhood mortality, and life expectancy didn't change much until very recently. In mid-nineteenth-century Bavaria, for example, women still had an average of eight to ten children — about one every 18 to 24 months until about the age of 40, if they lived that long. The interval between children was a bit longer if the previous baby survived, a bit shorter if it didn't. Life expectancy remained about 30 years in Greek and Roman times and was still only 34 years in eighteenth-century Sweden. It rose to about 40 years in the United States in 1850 and to 41 years in Britain in 1870.

But with improvements in standard of living, nutrition, sanitation, and safety, life expectancy began to increase dramatically. For example, life expectancy in Britain rose from 41 years to 53 years from 1870 to 1920 — four times the three-year improvement seen in the prior 50 years.

Just since 1970, worldwide life expectancy has increased to

the point where a baby born today can be expected to live to about age 71 to 72 years. The worldwide percentage of deaths occurring in the first five years of life has declined from nearly 25 percent in 1990 to just half that rate by 2013. Nearly 25 percent of deaths now occur in people over age 80.

The longest life expectancies are among Icelandic men, at 80 years, and Japanese women, at 86 years. Life expectancy in the United States is slightly shorter, but most of that's attributable to higher infant mortality rates. Life expectancy after age 65 is essentially the same in the United States as in other highly developed countries. Even at age 80, we can expect to live an average of almost eight more years.

Not surprisingly, we not only live longer, we're also healthier — as evidenced by the fact that we're getting taller. In western Europe, for example, the average increase in height went from less than half an inch in the nineteenth century to more than three inches in the twentieth century.

Increases in life expectancy go hand in hand with changes in worldwide causes of death and disability. Just since 1990, malnutrition has fallen from the 11th leading cause of death worldwide to the 21st, and the disability burden of all nutritional deficiencies worldwide has declined by about 40 percent. At the same time, disability caused by diabetes has increased by 30 percent, and diabetes has risen from the 15th to the ninth leading cause of death. Overweight, diabetes, and prediabetic elevations in blood sugar *each* cause more deaths worldwide than malnutrition, and *combined* they account for more than seven times as many deaths!

Since 1990, heart disease caused by high blood pressure has increased by about 6 percent worldwide, and high blood pressure now causes more global deaths than any other risk factor. By comparison, deaths and disability from diarrhea, perhaps the leading cause of life-threatening dehydration, declined by more than

60 percent. Excess salt intake, part of our protective mechanism against dehydration, now causes more than twice as many deaths as diarrhea. Also since 1990, worldwide deaths from injuries have declined by 13 percent, while disability caused by depression is up more than 5 percent.

Worldwide trends are informative, but how about the United States? The domestic news also isn't so good. Obesity, diabetes, high blood pressure, and depression collectively are already more of a problem here than they are in much of the rest of the world, and we're not seeing a lot of progress. For example, men and boys have continued to get fatter over the past decade. Weight finally appears to be stabilizing in women and girls, but it's stabilizing at a terrible level — obesity rates in adults have more than doubled since about 1970, and they have more than tripled in children and adolescents. Diabetes rates also are rising and causing progressively more deaths.

Our prevalence of high blood pressure has increased by about 5 percent over the past 10–20 years, largely because of increased obesity. Rates of successful suicide are pretty stable, but only because of outstanding emergency care for the increased number of suicide attempts. Successful suicide is the sixth leading cause of lost years of life, and depression is the second leading contributor to years lived with disability.

The only good news is that the risk of dying from a heart attack or stroke has fallen dramatically in the United States over the past 50 years, thanks to a combination of better prevention — lower cholesterol levels, less smoking, better treatment of blood pressure — and marked advances in medical and surgical treatments. But heart attacks and strokes still rank first and fourth among causes of death in the United States (and first and second globally) and dwarf the risk of bleeding to death.

Things have already moved in the wrong direction in poorly educated American adults. In 2010, Americans in well-educated parts of our country — such as Marin County in California and

Fairfax County in Virginia—lived as long as anywhere in the world. But largely because of their overprotective traits, Americans in poorly educated counties—such as Perry County in Kentucky and McDowell County in West Virginia—don't live any longer than people in Bangladesh.

And other parts of the world may not be far behind. Both India and China already have more diabetics than the United States. And by 2035, the number of diabetics will triple in India and increase about fivefold in China.

LOOKING TO THE FUTURE

By all projections, worldwide childhood mortality in 2030 will be only about half what it was at the beginning of the twenty-first century. This decline will drive a further increase in overall life expectancy to a projected 85 years in women and 80 years in men in high-income countries. The major outlier is sub-Saharan Africa, where life expectancy will still likely be about 55–60 years, even in 2030.

In 2030, scientists predict that heart attacks and strokes—diseases of clotting—will remain the two leading causes of worldwide death. Diabetes, related overwhelmingly to obesity, is projected to rise further, from ninth to seventh, in part because the current childhood and adolescent obesity rates portend a further increase in adult obesity. Diseases caused by high blood pressure will also increase, with blood pressure–related heart disease rising from 13th to 11th. And kidney disease, related both to high blood pressure and diabetes, will rise from 17th to 13th. Suicides will likely increase from 14th to 12th worldwide, higher than murder and war fatalities combined. And if we look at quality of life, including both years of life lost to premature death as well as how much we enjoy those years, depression will become the second leading problem worldwide and by far the leading problem in high-income countries such as the United States.

The other conditions projected to increase over time are mostly related to wear and tear on our aging bodies. For example, leading causes of disability in high-income countries by 2030 will include Alzheimer's disease and other dementias, osteoarthritis, and adult-onset hearing loss. Another increasing cause of disability and death will be cancer, which becomes more common as we age. The leading underlying causes of cancer include environmental toxins (such as tobacco smoke), infectious diseases (such as the human papilloma virus, which causes cervical cancer and a good portion of head and neck cancers, and hepatitis viruses, which cause liver cancer), and obesity, which is thought to account for about 17 percent of US cancers.

CAN OUR GENES TURN THE TIDE?

Since natural selection spreads beneficial mutations and eliminates harmful ones, could the same natural selection process that proliferated our current overprotective traits now eliminate them? Or might they be offset or counterbalanced by new mutations that are better for our current conditions? Either way, wouldn't it be wonderful if these adverse traits just disappeared on their own? Unfortunately, that's unlikely to happen. Let's consider why.

Eliminating Counterproductive Traits

Natural selection favors any trait that increases reproductive success early in life, even if it's bad for us later in life. That leads us to the following problem: the overprotective genes that drive us to eat too much food, consume too much salt, get too sad, and clot too much currently don't kill us quickly, don't prohibit us from finding a mate or having children, and don't prohibit those children from living long enough to have children of their own. But could they worsen to such a degree in the future that they could be vulnerable to natural selection?

For two of our overprotective traits — high blood pressure and clotting — we probably won't see any major effects on child-bearing or fathering, because these adverse effects rarely limit growth, development, or reproduction, especially in modern societies with currently available health care. How about anxiety and depression? People with depression and anxiety, especially if the conditions are severe, may be less likely to find mates and reproduce. But, as noted in chapter 4, sadness and depression have existed throughout recorded human history. Depression, at least of a limited duration, seems to be an advantageous adaptation that helps us abandon unrealistic ambitions and adapt to our status within a group. And, also as noted in chapter 4, anxiety is part of our early warning system, so it, too, is unlikely to disappear.

Could obesity and diabetes be a different story? Diabetes increases the risk of miscarriage by 30–60 percent — so women with normal blood sugar levels have perhaps a 45 percent repro-ductive survival advantage. If nondiabetic women have more off-spring, and if their children are less predisposed to develop diabetes, the percentage of nondiabetic people would gradually rise. However, as noted in chapter 1, calculations regarding the spread of beneficial mutations generally assume that all people — those with and without beneficial or harmful mutations — reproduce at equivalent rates. Most women in developed countries now marry later in life, control their own reproduction, and typically limit their number of pregnancies by using various methods to prevent them. Small differences in their reproductive choices can dwarf the impact of a substantial genetic survival advantage or disadvantage.

Another possibility is that obesity and diabetes will become so severe that obese diabetic people can't stay above the threshold needed to perpetuate themselves, especially as people in devel-oped countries marry later in life, when they may already suffer their adverse health consequences. Obese American women have generally been only 40 percent as likely to have children and less than half as likely to have more than one child compared with normal-weight women, in part because of irregular ovulation and the fact that their uteruses are less receptive to fertilized eggs. Obese mothers also are about 30 percent more likely to have a miscarriage, 20 percent more likely to have a premature delivery, and about 30 percent more likely to have their children die before adulthood.

However, recent data suggest that as obesity has become common and accepted, and as medical care has improved, obese women may no longer lag very far behind normal-weight women in the number of babies they bear. When you add up all the data, the offspring of obese and even type-2 diabetic mothers generally live sufficiently long to have enough children of their own, so natural selection is unlikely to eliminate obesity and diabetes. A more likely scenario is that obese people will become as obese as they can be while still being able to have children who also survive. And when obese people mate with other obese people, their surviving children tend to perpetuate the unhealthy cycle because they're about 14 times more likely to be obese than are the offspring of two normal-weight people.

Proliferating New Mutations

Now let's look at the second option—new mutations that could offset our overprotective traits. As we consider this possibility, it's important to recognize that natural selection is only one of three ways that mutations can be passed down to subsequent generations. The other two processes are *gene flow,* which is the movement of genes into or out of a local population by migrating individuals or groups, and *genetic drift,* by which random events may perpetuate and spread a mutation or eliminate it.

Perhaps the best example of gene flow is the so-called founder effect, in which a prolific male and his many children move to a previously unsettled location and, in essence, bring all or nearly all their genes. With subsequent inbreeding, these genes—including any mutations they brought with them—are perpetuated in their descendants as long as they don't confer disadvantages that wipe out the entire settlement. Genes also flow into a population when outsiders, especially men, move in and mate with local women—just as Viking men brought hemochromatosis to northern Europe, Genghis Khan brought his Y chromosome to central Asia, and about 60 percent of Y chromosomes in Mexico City are of Spanish origin.

From a man's perspective, there's no theoretical limit to how many women he can inseminate or how many children he can have. Monogamy can focus a man's attention on protecting a limited number of offspring. By comparison, polygamy will generate more offspring, but they may be more vulnerable because they'll receive less fatherly protection. Of course, if some men are polygamous, other men have fewer options. Healthy, strong men can and historically often have inseminated a disproportionate share of women and, as a result, tended to pass along the genes that gave them these advantageous traits.

As the human population has grown, there are now many more of us who have the 65 or so new mutations that, as I described in chapter 1, each of us has when compared with our parents. Since the human population has increased from roughly a million people 10,000 years ago to seven billion today, that's 7,000 times as many people with 65 or so random new mutations in each generation. Even though some of us may have the same random mutations as others, the overall number of mutations across the world is exploding, just as the population is exploding. Scientists estimate that more than 10,000 human mutations have been perpetuated by natural selection in the past 80,000 years. But what's interesting is that the mutations include 3,000 or so each in Africans, Europeans, Han Chinese, and Japanese, with little overlap among the four groups. This phenomenon isn't so surprising—we live in such diverse parts of the world that some mutations may be beneficial in a local niche, even though they're not helpful more broadly. For example, indigenous Siberians have not one but three adaptive mutations that, compared with the rest of us, help them metabolize fat into heat more efficiently and reduce heat loss through the skin.

When a new mutation first appears, it faces a daunting challenge because its perpetuation is dependent on a single person who can only give it to 50 percent of her offspring. In these early stages, the mutation's fate is as vulnerable as the multigenerational success of its initially very small number of carriers. The greater its survival advantage, the more likely it is to break out from its initial source and go viral. The mathematical likelihood that a single advantageous mutation will ultimately spread to the entire population is two times its percent benefit—so a mutation with a 10 percent survival advantage has a 20 percent chance of eventually spreading to every human being.

In genetic drift, a neutral mutation, defined as one that has no effect on human procreation or survival—such as the mutation that causes about 25 percent of us to sneeze when we're suddenly exposed to bright sunlight—is simply subject to the laws of chance. If someone who inherited a neutral mutation by chance happens to have a lot of children, the mutation will spread. But in small groups of humans, such a mutation can just as easily disappear. Since these random, neutral mutations arise and disappear at reasonably predictable rates, their accumulation over time can serve as a molecular clock that, like rings in a tree, can tell us how old our species is.

One of the best examples of the uses of a molecular clock is the estimate that all humans started with a single maternal ancestor—often called mitochondrial Eve—who probably lived somewhere in East Africa about 120,000 to 200,000 years ago. How do we "know" this? Although we get 23 of our 46 chromosomes from each parent, we get all the DNA in our mitochondria—which act as small power plants within our cells—from our mothers. Since we know that mitochondrial DNA accumulates a neutral mutation about once every 3,500 years, we can add up the mutations and estimate when, back in time, we all came from a single mother.

Using a similar approach based on neutral mutations in the male Y chromosome, we all seem to have had a common father somewhere around the same time. But these are just estimates, and they don't by any means either prove or disprove whether mitochondrial Eve and Y-chromosome Adam ever knew each other.

Random neutral mutations—genetic drift—accumulate over time, and the human species now has a huge number of them. For example, humans have anywhere from 10 to 40 million *polymorphisms*—which occur when more than 1 percent of people share an apparently neutral mutation. Each of us has about three million of these polymorphisms, more than 10,000 of which affect protein coding and about 1,500 of which alter the subsequent function of that protein. This accumulation of neutral polymorphisms provides a potential reservoir of mutations that might be helpful if and when things change in the future.

Sometimes an apparent polymorphism actually isn't as entirely neutral as the sneeze reflex or a dimpled chin. If it has both some advantages and some disadvantages, like the *DRD4* mutation, which predisposes us to risk taking, its frequency may reach a relatively stable middle ground in the population, consistent with both its benefits and hazards. This situation, called *balanced selection,* is seen in a number of mutations that make our red blood cells relatively resistant to the malaria parasite but also cause varying degrees of dysfunction of the red blood cells themselves. The balance for or against such a mutation could change rapidly if, for example, the risk of malaria increased or decreased.

In addition, some mutations that are truly neutral now may

have been perpetuated because they were beneficial sometime in the past. I mentioned this possibility in relation to the CCR5 protein and HIV back in chapter 1, but a more definitive example of this phenomenon is *mad cow disease* in humans.

Mad cow disease is the lay term for what scientists call bovine spongiform encephalopathy, an infectious form of Creutzfeldt-Jakob disease (CJD). CJD is a rare neurodegenerative condition associated with abnormal movements, a rapid decline in mental function, and death soon thereafter. Some cases are genetic and run in families, while a second type appears sporadically for no apparent reason. The third type is infectious CJD, which has been caused by receiving contaminated blood, corneal transplants, or growth-hormone injections used for short-stature children. But when we think of infectious CJD, we usually think of mad cow disease (also known as variant CJD), which has developed in several hundred humans who ate the meat of sick cattle.

The most dramatic example of apparently infectious CJD first appeared in the 1920s in the Fore tribe of Papua New Guinea, where the disease was known as *kuru*. The Fore had an interesting custom of ritualistically eating their dead elders—seemingly as a sign of respect, not of food shortages. Kuru developed only in women and children, and only in those who ate the brains of their elders.

Scientists quickly assumed that kuru must be infectious, since the brains of dead elders seemed to be a clear common source, but initial experiments showed no evidence of any bacteria, viruses, parasites, or other infectious agents. When I was in medical school, the common wisdom was that kuru must be caused by some undetected, very slow-acting virus, which was simply hard to find.

But my former San Francisco colleague and Nobel Prize winner Stanley Prusiner defied conventional wisdom by insisting that there really was no live infectious agent. He showed that the disease was transmitted by a totally unique mechanism—proteins that were misfolded and then induced other proteins, like a sequence of dominoes, to misfold in the same way. The protein of interest, called a *prion,* is just one of innumerable proteins in the brain, and its function still isn't entirely clear. The prion protein normally exists in a helical three-dimensional structure, and this structure is critical for its normal function and interaction with other proteins in the brain. But the same sequence of amino acids that make up this helical shape can also line up differently in three-dimensional space to form a relatively flat sheet. In this flat-sheet configuration, the prion protein not only may not perform its normal function, it also tends to aggregate and form clumps that literally gum up the works in the surrounding brain tissue.

But it turns out that a relatively common genetic polymorphism called M129V alters the prion protein to make it less susceptible to the misfolding cascade in kuru and in mad cow disease. For reasons we don't understand, this polymorphism is present in a widely varying percentage of people around the world, ranging from a low of 1 percent to as high as 50 percent. But now—within just 90 years after kuru's first known appearance in the Fore— it's found in nearly 90 percent of Fore women, presumably because those without it died from kuru. Interestingly, this polymorphism isn't totally protective, but it can delay the onset of kuru by decades—long enough for Fore women to reproduce and rear children. At least so far, no one worldwide with this polymorphism has yet to be diagnosed with mad cow disease.

But just to show how quickly we can spread protective genes in the face of a truly lethal challenge, about half the potentially susceptible Fore women without this M129V polymorphism have a new mutation in the prion gene called G127V. This mutation, which is nearly adjacent to the M129V mutation, also protects against kuru and hasn't been found anywhere else in the world.

The widely varying frequency of the M129V polymorphism throughout much of the world raises another question. Perhaps it's not a truly random polymorphism but rather a mutation that, sometime in the past, protected our brain-eating Paleolithic ancestors from a prehistoric version of mad animal disease. Such a hypothesis would be consistent with the finding that no humans have ever contracted infectious CJD from sick sheep who suffer from a prion disease called scrapie, apparently because all humans share a gene that protects us from getting "mad sheep" disease.

As the human population has expanded, the number of mutations has gone up as well. Just by chance, a rare mutation that might have appeared once every 50 or more generations in the Paleolithic era could now appear somewhere in the world within one generation. Of course, if it took 7,000 times as long for a mutation to spread to a population that's now 7,000 times larger, then the more frequent appearance of mutations wouldn't have any more impact on the human population.

But remember, as we saw in chapter 1, that's not what happens — mutations spread slowly at first and then more rapidly. The enlarging worldwide reservoir of neutral mutations is waiting around for possible natural selection. If and when one suddenly becomes advantageous, it would, on average, take less than twice as long to spread through the predicted eight billion humans in 2030 as it would have to spread through one million people 12,000 years ago.

These theoretical calculations assume that we all come into contact with one another. Modern intercontinental travel and trade make exponential spread reasonably realistic for a highly contagious infectious disease, but the same very rapid spread can't be assumed for genetic mutations, which require humans to mate and have children together.

If a new advantageous mutation confers an extraordinary benefit — perhaps because it saves us from an infectious disease or environmental disaster that kills most other people — it's easy to imagine how the few worldwide survivors would interbreed and spread it rapidly. Back in chapter 1, we learned that HIV could theoretically have killed all but the small percentage of us with two *CCR5* mutations if it had spread throughout the world in an era before antiviral medications. Although we can't be sure, prior challenges that killed a large percentage of humans may similarly have left us with just the survivors who had advantageous mutations. For example, plague (also called the Black Death) probably reduced the European population by 50 percent or so between the sixth and eighth centuries and then by about one-third when it recurred during the fourteenth century. And when the Europeans invaded North America, as many as 90 percent of the native population may have died from its first exposure to influenza, measles, and smallpox. Maybe many of the survivors of plague, smallpox, and other infectious catastrophes were just lucky, but it's more likely that many of them benefited from a mutation that previously gave them little or no advantage but suddenly became truly lifesaving.

Although some mutations may be truly lifesaving, many of the mutations I've described in prior chapters provide *relative* benefit — perhaps you would have three surviving children rather than two — but were not absolutely required for survival. Relatively beneficial mutations can't spread that fast — people without them don't all die off, so the beneficial mutations just get progressively more common as they spread.

To get a sense of how rapidly a relatively beneficial mutation can spread, let's consider a hypothetical example. Imagine that having two copies of a mutation, one from each parent, is better than having only one, yet one is still better than having none. Then let's assume people with two copies of the gene do twice as well as people with none, and people with one copy are halfway in between. Even if this mutation started off as a very common polymorphism in 10 percent of the population, it would take five generations to get to about 40 percent of the population. Five generations may be a blink of an eye in the 200,000-year history of the human species, but it's five times as long as the one generation between the first report of AIDS as a clinical syndrome and the development of effective medications to treat it.

So we shouldn't get carried away — it's true that the current size of the human population means that a random advantageous mutation is 7,000 times more likely to arise now compared with 10,000 years ago and could theoretically spread to the current population about half as fast. But it's not realistic to assume that every human would get the mutation, simply because we're now so numerous and geographically dispersed that all humans no longer routinely interbreed with one another. The bottom line is that we can't expect new mutations or the rapid spread of any existing mutations to reduce the frequency of obesity, high blood pressure, depression, and excess clotting.

We also have to realize that the accumulation of initially neutral mutations is a double-edged sword. Just as some of them may be serendipitously advantageous in the future, some of them become serendipitously disadvantageous. And time allows disadvantageous mutations to settle in. For example, consider a mutation such as Tay-Sachs, which first arose about 1,500 years ago and doesn't cause any problems if we get it from only one parent but is uniformly fatal if we get it from both. Because people with just one copy of the mutated gene do fine, the mutation never goes away, even though people with two copies of it never have any children. Furthermore, new spontaneous adverse mutations will add to the pool. Doing the arithmetic, we can calculate that a mutation like this, which arises in one in 10,000 people, will ultimately reach a steady-state equilibrium so that about 2 percent of the population will have one copy of it, even if every person with two copies of it dies. Interestingly, scientists estimate that about 30 percent of modern humans have at least one such mutation — a single copy of a mutated gene that would be lethal if their offspring inherited it as well as a second mutated version from their other parent.

CAN THE ENVIRONMENT CHANGE US FASTER?

Although our genes govern who we are, our bodies can also adapt to the environment as we grow and develop. For example, as children grow, the muscles in their eyes strengthen depending on what they use their eyes to see — and these muscles influence the shape of our eyeballs. In industrialized countries, a substantial proportion of children are nearsighted, a trait that would have been devastating to our hunter-gatherer ancestors, who needed good distance vision to avoid all the hazards of prehistoric life. Nearsightedness certainly has a genetic component, but it seems to be overwhelmingly related to the fact that children now spend less time outdoors in the bright sunlight and more time indoors with artificial light, which is dimmer than natural sunlight. By exercising their eye muscles for indoor and up-close vision, as many as 80 percent of adolescents may have some degree of nearsightedness, and the rate is highest in people who are the most educated. By comparison, just 40 extra minutes of outdoor time daily can reduce the development of childhood nearsightedness by 25 percent.

Of course, nearsightedness isn't killing us and is unlikely to affect the future of human survival. No one really thinks that even seriously nearsighted people will have trouble functioning, finding mates, or reproducing, especially if they wear corrective lenses. But nearsightedness is an example of how an individual's growth adapts to changes in the current environment.

In the future, are such adaptations likely to help? Or are they more likely to reduce our fitness and perhaps even our reproductive success? It's hard to know for sure. If our muscles somehow become less efficient and need more calories to power them, rates of obesity could decline. But if humans gradually have less muscle mass because we get progressively less active — a much more likely scenario — the reduction in muscle mass will exacerbate our caloric excess, further reduce our ability to burn calories, and induce us to get even fatter. So if adaptations generally tend to reinforce behavior—including behaviors that are already making historic survival traits counterproductive—they'll make us less fit, not more fit.

It's one thing to consider how our environmental experiences might change us, but how about our parents' experiences? Could we somehow pass down what our bodies have experienced, just as we pass down parental wisdom? Could our lifetime experiences somehow get into our children's genes?

Well before Gregor Mendel's work with peas jump-started our knowledge of genetics, and before Charles Darwin described his theories of natural selection, Jean-Baptiste Lamarck of France proposed that we inherit traits from our parents. But his idea of inheritance was very different from Mendel's and Darwin's. He believed we're a product of our environment: if we use an organ frequently, we'll strengthen and enlarge it; if we use it infrequently, it could disappear. Although these beliefs don't seem so different from what explains the increase in nearsightedness, Lamarck also made a bolder statement: he believed that use-related changes in one generation would be passed on to the next.

Lamarck's most famous example was the giraffe. He postulated that giraffes' necks gradually got longer, generation after generation, because they had to stretch more to reach leaves high in trees. If enough leaves were available in short trees, he believed a giraffe's neck wouldn't grow as long. Furthermore, he argued that a young giraffe would inherit this experience-derived trait and grow up to have a neck about the same length as its parents.

Lamarck's theories were taken to a calamitous extreme by Trofim Lysenko, a Soviet agronomist who rose to prominence as director of the V. I. Lenin All-Union Academy of Agricultural Sciences. Lysenko claimed that exposing a wheat seed to extremely low temperatures would make it hardier and that it would pass on that adaptation to subsequent generations. Joseph Stalin, hoping this approach would yield cold-resistant wheat, embraced Lysenko's teachings, abandoned genetic research, and ostracized Soviet geneticists. The results, of course, were disastrous — a huge investment in agriculture failed, and the Soviet Union found itself an importer rather than exporter of grain. Lamarckian theory fell into disrepute as plant genetics spurred huge increases in disease-resistant crop production in America and western Europe.

We are a combination of our genes at birth — *nature* — as well as all the things we have experienced since we were in the womb — *nurture*. Lamarck's error, taken to an illogical extreme by Lysenko, was thinking that most or all of a person's nurture experiences could be passed on to the next generation.

But it turns out that Lamarck wasn't all wrong. Recently, scientists found that our genes can be influenced by environmental factors that can differentially tag our DNA and affect how genes are expressed by causing *epigenetic* changes. If our genes represent our nature, these tags represent nurture.

From the time we're in the womb until the time we die, our DNA can get tagged or untagged in response to things we experience. This process partially explains how we grow, develop, and get sick. For example, exercise induces epigenetic tags that influence the function of both our skeletal muscles and our fat tissue. Epigenetic tags also explain why identical twins reared in different environments don't act quite as alike as twins who are raised together.

To understand why our genes get tagged, let's go back to the egg and sperm, each of which carries 50 percent of a baby's chromosomes and genes. When an embryo is first formed, it has just one cell — with half its chromosomes and DNA from the egg of the mother and half from the sperm of the father. This one cell has to be a *totipotent* stem cell — able to make every cell in a newborn's body. As the fetus develops, the genes are copied into each of our approximately ten trillion cells. This one cell also has to be sure that each of those subsequent cells can do what they need to do as the baby's body grows to full maturity, at least long enough to perpetuate the species.

How can one person's DNA make a brain cell in the head, a muscle cell in the arm, a bone cell in the skeleton, and a red blood cell in the bone marrow within that skeleton? Although each of our cells could be making 21,000 proteins continuously from its 21,000 protein-coding genes, that's not what happens. Every gene is dressed up and ready to go, but only about 10–20 percent of them get activated in each cell, and it's this set of active genes that defines what the cell becomes. For example, the identical gene that churns away making a protein needed in a muscle cell may be on temporary or even permanent sabbatical in a brain cell. And as an embryo's organs develop, the cells that comprise them are directed by epigenetic tags that get placed on their DNA and that tell them whether they should become heart cells, bone cells, liver cells, or another type of cell. As a result of modern stem cell research, we now know some of the local signals that apparently cause cells to be directed to specific fates, but we still have a lot to learn. For this process to work, the embryo's original totipotent cell has to be essentially devoid of epigenetic tags, but future generations of embryonic cells must become progressively tagged as they go about the business of differentiating into specific types of cells and organs.

This embryonic tagging process shows that nature and nurture really aren't that different—nature is the genes in our first cell, while the environment (nurture) tags and modifies our genes thereafter. It also raises the possibility that if embryonic cells can pass on their epigenetic tags in response to a changing local environment during fetal development, perhaps the human species might change our genome more rapidly if we could pass down the epigenetic tags we accumulated during our lifetimes.

Scientists originally believed that all the epigenetic tags accumulated during a person's life got erased when men formed their sperm (throughout their adult lives) and when women formed their eggs (mostly when they themselves were in utero but also, as we now know, to some extent during their reproductive lives). After all, our sperm and eggs can't carry the epigenetic tags that tell cells what organs they should form—otherwise that first fetal cell couldn't be totipotent. We now know, however, that not all epigenetic tags get erased when sperm and eggs are formed— some that don't affect organ development actually make it to the embryo. These tags almost always disappear by about two weeks after the egg and sperm form the first totipotent cell. But even the very small percentage of persistent tags means that we may inherit hundreds or even several thousand of them across at least one generation.

The occasional persistent epigenetic tag may explain how a child's behavior sometimes can be linked to experiences faced by one or both of their parents before the child was conceived. For example, a father's obesity can affect the tagging of his sperm, and animal experiments have demonstrated that those tags can influence how a father's offspring metabolize sugar.

Evidence for epigenetic inheritance has been shown for eight generations in plants and at least two generations in fruit flies, worms, and mice, but it's less clear whether epigenetic tags get passed on to subsequent generations of humans. However, epidemiological data suggest that a propensity toward obesity and diabetes can be passed down from poorly nourished grandfathers to grandsons and from poorly nourished grandmothers to granddaughters, implying but not proving a sex-linked epigenetic effect.

Could epigenetic inheritance help us adapt more quickly to our modern environment and address our overprotective traits? Perhaps, but right now it's speculative at best. For example, detailed studies in Quebec looked at death records among people who were conceived and born in urban and rural locations in the seventeenth and eighteenth centuries and then either remained in the same type of location or moved from rural to urban or urban to rural. If epigenetics provided a major survival advantage in an era when about one-third of children died by age 15, you might expect that those who were born and lived in the same type of locality — whether rural or urban — could have inherited some adaptive advantage from their parents compared with those who were born in one type of locality and then migrated into the opposite type. But scientists so far have found no evidence for such a benefit.

We must also remember that although epigenetic tags can be beneficial, they can also be detrimental. The first environmental exposures that can cause epigenetic changes occur when we're still in the womb, and we know that this uterine environment can negatively influence the baby. A mother's malnutrition during the first trimester of pregnancy, for example, has long been recognized as a factor in her offspring's higher likelihood of developing adult obesity, and children of mothers who smoke during pregnancy are 30 percent more likely to become obese. The likely explanation is the thrifty phenotype discussed in chapter 2. But also as discussed in chapter 2, more maternal weight gain likewise increases childhood obesity, presumably via another set of epigenetic alterations.

Furthermore, many of the epigenetic changes that we accumulate during our lifetimes result from damage by environmental toxins. And these epigenetic tags appear to be one of the major causes of cancer.

What this means is that any epigenetic inheritance that may exist is, like the initially neutral gene mutations previously discussed, a double-edged sword. It's possible that we may inherit some things that could help us change in a beneficial way, but we'll probably also inherit some epigenetic changes that are deleterious. Over time, natural selection for epigenetic as well as genetic changes could have a big effect on our species. But we shouldn't expect it to lead to immediate, highly beneficial, and widespread changes in the human species at the same rate that our lifestyle has changed since the advent of the Industrial Revolution.

Natural selection is a wonderful process: it's helped us adapt over millennia, and it's probably accelerating. But even the accelerating speed of natural selection—and even if it's driven by epigenetics as well as genetics—can't possibly match the rate at which our world has changed and continues to change. If we're going to blunt or reverse the traits that make us eat more than we need, crave salt, get too anxious and depressed, and clot too much, we can't count on our bodies to do it for us naturally. Instead, we'll either have to find ways to change our behavior—mind over matter—or we'll have to rely on science and medicine to help us.

CHAPTER 7

Changing Our Behavior

Oprah Winfrey, worried about her increasing weight and its adverse effects on her health, demonstrated her motivation and determination when she successfully lost 67 pounds — going from 212 to 145 — in 1988. After she quickly regained that weight and more, she lost around 85 pounds — going from approximately 235 to about 150 — in the 1990s. In 2005, subsequent weight gain motivated her to starve herself for four months to get down to 160 pounds. But by 2008, she had regained 40 pounds, and her weight has continued to yo-yo up and down ever since. Although Oprah is unique in many ways, her weight problems are all too common — about 90 percent of overweight people ultimately regain most or all of the weight they lose. Oprah Winfrey's struggles to achieve a healthier weight are emblematic of how hard it is for us to change our habits. And we face similar difficulties regardless of whether those habits and behavioral responses were so critical for survival that they're built into every human's genes or whether (like the mutation for adventurous behavior discussed in chapter 2) only some of us have them.

*　　　*　　　*

Each of the four key survival traits we've covered—eating more food than we need, craving salt to protect our blood pressure, being at risk of getting anxious or depressed, and clotting too much—is deeply rooted in the genes we inherited from our ancestors. And as I mentioned in chapters 1–5, some of us have specific mutations that make these traits even stronger. But all of us now have a panoply of protective genes that allowed our ancestors to win countless battles of survival for more than 200,000 years.

These genes continue to exert powerful influences on our behavior and our basic bodily functions, regardless of whether we really need them as much as we used to. And although our DNA will continue to mutate as we and our descendants face new challenges, it can't possibly change fast enough to keep pace with the remarkable progress of civilization.

So let's ask a different question: Can we harness our willpower to change our behavior and thereby offset our genetic tendencies? Is "mind over matter" a reasonable approach to bridging the disconnect between our historic survival traits and our current best interests? Or are we so programmed, much like salmon returning to their birthplace to spawn and die, that behavior change is hopeless?

Numerous aphorisms—"The road to hell is paved with good intentions"; "The best-laid plans of mice and men often go awry"—underscore how difficult it is for us to change our long-held behaviors. Simply put, if it were easy for us to change bad behaviors, none of us would have any of them.

In reality, we all have habits we would like to change. Each year, as many as 50 percent of us make one or more New Year's resolutions—losing weight and increasing exercise are among the most common. And although many of the behaviors that we would like to change are literally part of our DNA, none of our habits is absolutely set in stone. Oprah has certainly struggled to keep her weight down, but even her intermittent success gives us some hope. Sometimes we can and do change.

Good plans are a prerequisite to changing behavior, and some data suggest that we can actually be taught to plan better. We're substantially more likely to succeed if our goals are accompanied by what are called *implementation intentions,* which specify our plans for initiating a behavioral change, protecting ourselves from distractions, abandoning unsuccessful strategies while finding new ones, and conserving energy and enthusiasm so as to be able to continue for the long haul. Our good intentions for change are also more likely to become reality if we include step-by-step ways to monitor our progress.

But actually implementing our plans also requires what is sometimes called *executive control,* which includes not only good planning but also flexibility in finding ways to solve unexpected problems that may arise, focusing on specific tasks that may be necessary to achieve the goal, and learning to suppress automatic or reactive behaviors that will interfere with the goal. We also can do somewhat better if we put some money at stake, combine a difficult behavior change with something we like (for example, if we exercise while watching a favorite television show), and join a group that is trying to make similar changes.

And in achieving these goals, "perfect" may be the enemy of "good." Data suggest that perfectionism may be counterproductive, since perfectionists may abandon a goal if they can't do it perfectly, whereas the rest of us may have more flexibility in altering plans or strategies to get as close to the goal as possible.

Unfortunately, failure is the norm. Even with the best implementation plans, we're still much better at reinforcing prior behaviors than at truly changing behaviors. And the stronger the habit, the harder it is to break, regardless of how good our planning, implementation intentions, and executive control may be.

Why do we usually fail? First, many of us never even get started, and the longer the delay between when we set a goal and when we actually begin to strive for it, the less likely we are to get there. Some of us may initially make progress but then get derailed when familiar situations and old friends make our prior bad behaviors seem comfortable again. Because habitual behaviors are often triggered by situational cues—for example, we're more likely to overeat when we see others overeating—our expressed intentions are poor predictors of actual behavioral change unless the triggering situation can be changed or the habitual and sometimes unconscious response can be altered.

Another problem is delayed gratification—most behavior changes don't immediately make us feel better, look better, or be better received by our friends. In some situations, we may even get just as much positive reinforcement from others by declaring a goal—"I'm going to exercise more" or "I've started a new diet"— as we get from the behavior change itself, especially concerning outcomes such as weight loss, which may take months to become apparent to others. This premature sense of accomplishment can weaken our resolve to continue the uphill battle.

We also often have unrealistic expectations. We typically underestimate how much time and effort it will take to achieve a goal, despite our conscious recognition that previous goals have always been more difficult and time-consuming than we expected. To paraphrase Mark Twain, habit is habit, not to be flung out of the window but rather to be coaxed down the stairs a step at a time.

Finally, even the best implementation intentions struggle against our tendency to think it's fine if we occasionally succumb to temptation and indulge ourselves. Perhaps we think it's just a one-off exception for which we can later compensate, or we conclude that the indulgence is irresistible and only transiently available. Our remarkable and sometimes even subconscious tendency to justify self-indulgence is exemplified by an experiment in which people ate more chocolate in a taste test if they were first asked about justifications for personal indulgence in a seemingly unrelated questionnaire.

Despite our best hopes for change, our past behavior is about four times more likely to predict future behavior than are any new goals. As a result, we usually revert to our old behaviors, which stem from both nature and nurture — our genes and what we've experienced during our lifetimes. These behavioral realities challenge our ability to lose weight, reduce salt intake, increase exercise levels, and address anxieties and depression.

DIETING

This is not a diet book, so I'm not going to review the massive literature — medical or popular — on diets. The modern approach to weight loss emphasizes a variety of dietary regimens whose competing recommendations exemplify their disappointing success rates. The ballooning obesity epidemic itself is the best testimony to the dismal failure of the current behavioral approach to avoiding and treating overweight and obesity.

Back in chapter 2, I emphasized that it only takes about an extra ten calories per day for three years to gain a pound, so you might think it would be really easy to lose that pound. But let's face it: no one who wants to lose weight would be happy with losing just one pound in three years. The reality is that a sedentary 220-pound person wanting to lose 45 pounds in six months would have to reduce daily caloric intake from about 3,000 calories to about 1,800 calories. Or this same person could get about the same weight loss in a year by reducing daily intake to about 2,500 calories.

The problem is that dieting — any caloric restriction — isn't natural. We were made to eat, at least until we're full, sometimes uncomfortably so, and then to eat again as soon as we can. The only exception is that if we have access to only one food, we may get tired of it even before we're full—unless of course, we're at risk of starvation.

Because of all the hormones that kick in to try to keep us from losing weight, diets are hard to implement and even harder to maintain. Most weight-loss interventions, including counseling and dieting, show that people rarely lose more than an average of about ten pounds even after a year of trying. And comparisons among a variety of diets don't show very much difference, which of course explains why there are so many different dietary fads.

Although we're somewhat less efficient at absorbing calories from hard-to-digest sources (such as nuts) than from easy-to-digest sources (such as calorie-dense processed foods), it's how many calories you consume—not what food or foods they come from—that overwhelmingly determines how much you weigh. A low-fat diet causes more fat loss than does a low-carbohydrate, high-fat (Atkins or Paleolithic) diet. However, the Paleo diet has slightly outperformed the traditional low-fat diet in terms of actual weight loss in some (but by no means all) studies, perhaps because it has a different effect on the hormones that try to protect us from losing weight (which I discussed in chapter 2), or perhaps because people find it easier to follow. But the differences aren't very impressive. A low-carbohydrate diet, on average, results in at most a two-pound-larger initial weight loss and isn't clearly any better for keeping weight off once you've lost it. For weight control, what matters most is the ability to get enough satisfaction from a diet so that you'll actually follow it, and that's why no one diet is consistently superior to any other for weight loss in real life.

The only convincing exception to the principle that calories tell the whole story comes from data on mice who snack. In a careful experiment, mice who had access to a high-fat diet through-out the day didn't eat any more calories than those who had access to it for only eight hours a day. The only difference was that the mice who had constant access to food and ate throughout the day were able to spread the calories more evenly over 24 hours. These mice, whose eating pattern is the equivalent of frequent snacking, gained weight and, after 20 weeks, weighed almost one-third more than those who had access to food only eight hours per day. We don't currently know whether this difference is attributable to alterations in how food is absorbed from the intestine or how snacking may alter metabolism. Nevertheless, snacking and grazing throughout the day—which may have helped our hunter-gatherer ancestors maintain their weight during times of food shortages—now may contribute to the modern obesity epidemic.

A different and more complicated question is the overall health effects of various diets. Weight per se isn't the only issue. Some diets probably can make us healthier, even if they don't necessarily make us thinner. For example, a Mediterranean diet, which emphasizes olive oil, nuts, fresh fruits, vegetables, fish, legumes, white meat rather than red meat, and wine (while discouraging soda, commercial baked goods, trans fats, and processed foods), appears to reduce the risk of diabetes, better control existing diabetes, prevent heart attacks and strokes, and even perhaps slow the aging process by preserving the length of our chromosomal telomeres, out of proportion to the weight loss it may induce. By comparison, there's conflicting evidence and certainly no convincing verdict about whether a low-fat diet or a low-carbohydrate diet is preferred from a metabolic health perspective or in terms of preventing diabetes and cardiovascular disease.

But despite all the challenges and the frequent relapses, I'm certainly not arguing that we should ignore the potential benefits of a better diet. We should try to keep our saturated fat intake below 10 percent, consistent with the approximately 6 percent that scientists estimate our Paleolithic ancestors consumed. Overall fat intake, however, is a different story, since your cholesterol level won't rise appreciably with a higher intake of unsaturated fats, found in most fish and wild game. Protein intake can rise to 30 percent of calories without any harmful effects, as long as you don't have kidney disease. Carbohydrates currently make up close to 50 percent of our calories — and that's not a problem if it means 50 percent of the calories we actually need, but it becomes a big problem if it means 50 percent of too many calories. And there's reasonable evidence that we'll do better if more of the carbs come from fruits and vegetables, as in a Mediterranean diet, than from calorie-dense processed grains and sugars. But otherwise there's not much science to validate the popular belief that some carbs are better for you than others.

When we're trying to diet, a wide variety of behavioral therapy approaches — individual and group-based, in person, by telephone, and even over the Internet — can produce modest weight loss in motivated individuals. But, as with any weight-loss approach, maintaining the behavioral change is always difficult. Even very intensive lifestyle counseling, with food options individualized to a patient's preferences under the direction of a physician, results in an average of only about seven to ten pounds lost over the course of six months to two years.

The concept behind meal replacement diets — liquid shakes, meal bars, and prepackaged replacement plans — is that they're designed to satisfy as many of our natural instincts as possible. In short-term studies, programs such as Nutrisystem, Jenny Craig, and Weight Watchers have provided incremental benefit in the relatively small number of motivated people who enroll in them, with average weight losses sometimes as high as 15–25 pounds at the six-month mark. Some people have had truly remarkable and sustained benefits from these commercial weight-loss programs. But the cost of such programs and the required degree of commitment make it hard for most people to continue — so long-term results don't really look very different from those of other weight-loss approaches.

Back to Oprah and Yo-Yo Dieting

Oprah Winfrey was by no means the first—nor will she be the last—unsuccessful dieter. The term "yo-yo dieting," initially coined by public policy expert Kelly Brownell and his colleagues back in 1986, aptly describes how many people lose weight but then regain it, often almost as rapidly.

One of the most famous and best documented yo-yo dieters was William Howard Taft—the last in the sequence of obese American presidents discussed in chapter 2. In 1905, before Taft became president, he weighed 314 pounds and was so desperate to lose weight that he contacted Nathaniel Yorke-Davies, a London doctor and perhaps the world's foremost expert on obesity at the time. Taft followed Yorke-Davies's advice, and his detailed private journal chronicled about a 60-pound weight loss over the following five months or so. But the success was short-lived. By the time Taft was inaugurated as president, in 1909, his weight had ballooned to 354 pounds, 40 pounds more than he weighed when he initially contacted Dr. Yorke-Davies and about 100 pounds more than the lowest weight he had achieved. We don't have good records on Taft's subsequent weight, but there's no evidence that he got any thinner during his four years in the White House. In the year after his presidency ended, Taft's 70-pound weight loss was a front-page story in US newspapers. His weight went up and down after that, but he still weighed 280 pounds at the time of his death, in 1930, despite having been comatose for weeks before dying.

The struggles of Oprah and Taft are symbolic of a general phenomenon. In one survey, about 50 percent of northern European women, regardless of their weight, had made a New Year's resolution to lose weight in the previous two years. Among women who were obese, only 9 percent reported any success, whereas about 75 percent admitted to repeated failures. Interestingly, normal-weight women were about twice as successful at losing weight, suggesting that those of us who most need to lose weight are the least successful in doing so. Unfortunately, about one-third of all successful dieters will gain back at least 50 percent of their lost weight within one year. And yo-yo dieting probably makes it harder to lose weight in the future, perhaps because our bodies get even better at protecting us from weight loss.

Why We Fail

Although our genes undoubtedly influence our weight, our environment and habits explain more about differences in individuals' weight. Most diets fail because we go back to old habits in our old environment. Obesity is a chronic condition that requires a long-term approach designed to change our behavior while still addressing the variety and balance needed to satisfy our intrinsic food instincts.

One of the reasons we fail is that we delude ourselves. In the real world, many of us routinely lie — to ourselves and even to others — about how much we eat and how much we weigh. A detailed scientific analysis found that 40 years of self-reported American national survey data on exercise and caloric intake were "not physiologically plausible" in two-thirds of women and nearly 60 percent of men. Men underreported their intake by almost 300 calories per day and women underreported by about 350 calories per day. Furthermore, obese men and women were even less reliable, with obese men underreporting by more than 700 calories per day and obese women by more than 850 calories per day. Even William Howard Taft, who after his presidency became chief justice of the U.S. Supreme Court, lied to Dr. Yorke-Davies about what he really weighed.

Another challenge to staying the course on a weight-loss program is that the modern world gives us many choices — sometimes too many — that weren't available to our ancestors. Intrinsic choices that made sense in the Paleolithic era now cause us problems. A classic example is the size of a food serving. For our ancestors, it was easy—get the biggest portion you can find and eat it all before someone takes it away from you. Even now, we have the same instincts, which are well understood by movie theaters, where the portions of soda and popcorn range from large to humongous. But we don't complain. In one experiment, for example, moviegoers in a suburban Philadelphia theater were selected at random to receive free popcorn—either a medium-size container or a twice-as-large container. People who randomly received the larger container ate more popcorn than those who received the medium-size container—45 percent more when the popcorn was fresh and tasted good, and one-third more even when the popcorn was stale and tasted terrible.

In other studies, doubling the size of the container influenced people to eat about 20 percent more spaghetti and up to 50 percent more snack foods. Even experts in nutrition aren't immune to this phenomenon. In a fascinating experiment of nutrition experts invited to an ice cream social, the experts served themselves about 30 percent more ice cream if they were randomly given a larger bowl and about 15 percent more if they were randomly given a larger serving spoon—and nearly 60 percent more if they were given both! Controlling portion size really matters, and it can be problematic for us in today's world—whether it's the supersizing of food servings, larger containers, or the tendency to consume more if we snack directly from the package or eat ice cream directly from the carton.

The proximity and visibility of food also matter. In an experiment in which office secretaries were provided candy in a clear or opaque bowl, either on their desks or about six feet away, the secretaries ate about two more candies per day if the candies were in the clear bowl, about two more candies per day if the candies were on their desks, and about four more candies per day if the candies were easily visible in a clear glass bowl on their desks. Interestingly, the secretaries consistently overestimated how many candies they ate when the candies were six feet away and underestimated how many they ate when the candies were on their desks. This misperception is particularly interesting because, as noted earlier, reducing snacking may prove to be a way to avoid gaining weight— even if we consume about the same number of overall calories.

Our unconscious desire to join the "clean plate club" also presumably explains why people who are served soup in bowls that are imperceptibly refilled as the soup is eaten will eat almost 75 percent more soup than people who eat from regular soup bowls. We also tend to eat more, especially in restaurants, if the lights are low, the music is soft, we're eating with other people, we think it's time to eat, we see or smell food, or we're provided with a variety of food options.

Environmental influences on eating behavior (plate shapes, package sizes, lighting, layout, color, and convenience) increase food consumption because they either reset our sense of what's normal or make it harder to keep track of how much we're actually eating. These social cues lead to what nutritional scientist Brian Wansink termed "mindless eating"—subconscious tendencies that are hard if not impossible to change simply with education or data.

The food industry's goal is to sell more food, so it's not surprising that they have learned that we like calorie-dense food with a lot of sugar, salt, and fat. Since our stomachs sense fullness based principally on the volume of food we eat rather than on the number of calories we consume, we'll naturally consume more calories if we eat foods that have a high number of calories per volume. The food industry's combination of calorie-dense foods, large package sizes, and discounts on these large packages is a triple threat against maintaining a normal weight. What we really need are smaller bowls and plates, smaller portion sizes, smaller snack packages, smaller spoons, smaller ice cream cartons, fewer buffets, and timed locks on our food cabinets.

But even then, we're still influenced by peer pressure, and we look at others to appreciate social norms. If no one we know is obese, we're unlikely to want to be an outlier. But it also goes in the other direction. For example, if one of our friends becomes obese, our own likelihood of becoming obese goes up by more than 50 percent.

In an interesting experiment, women were randomly assigned to an intervention in which some of them were taught that model-like slimness was actually fake. They were shown pictures of supermodels in association with words such as *artificial, phony, bogus, sham,* and *false.* The women who were given these associations were significantly less likely to consider such slimness to be ideal and reported significantly higher levels of satisfaction with their own bodies, regardless of their weight.

Or consider Blair River, a 6-foot-8, 575-pound spokesperson for Heart Attack Grill, a restaurant known for its hefty hamburgers and fries cooked in lard. River was an unabashed advocate for the restaurant and its menu before dying at age 29. But even his death hasn't prevented people from continuing to order the restaurant's nearly 10,000-calorie "Quadruple Bypass Burgers," nor has it led the restaurant to change its policy of allowing people who weigh more than 350 pounds to eat for free. For people who don't want to lose weight or can't, the Heart Attack Grill exemplifies an unapologetic counterculture in which morbid obesity becomes acceptable and even glorified.

It's unlikely that we'll successfully fight the obesity epidemic if obese people don't feel bad about their inability to lose weight but instead accept or even celebrate their obesity. As overweight and obesity become more common, much of the social stigma goes away, and we have less incentive to try to fight our genes.

But we can also create interventional programs to change local norms. In one randomized study, women who lived in public housing in poor neighborhoods with their children were offered housing vouchers to be used if they moved to less poor neighborhoods as well as counseling on how to move. This opportunity led to a reduction in severe obesity and obesity-related diabetes in the families who moved, at least in part because of the local norms for lower body weight in the new neighborhoods, especially among children.

The Occasional Great Success

Every now and then, truly motivated individuals can defy the general statistics and our genes. Breanna Bond of Clovis, California, weighed 186 pounds at the age of nine. Not surprisingly, her parents initially thought her obesity must have been caused by a metabolic abnormality. But when all her tests were negative, Breanna and her parents began a diet and exercise routine that included a daily four-mile walk and 75 minutes on a treadmill. Within a year, she lost 66 pounds.

Another notable success is Jennifer Hudson, who won an Academy Award as the overweight member of the fictional singing group featured in the movie *Dream Girls*. After the birth of her son, the 5-foot-9-inch Hudson reportedly lost 80 pounds, going from about 220 pounds to 140 pounds — with her body mass index going from an obese 32.5 to a normal 20.5 and her dress size from a 16 to a 6. The secret to her success was no secret. She increased her exercise, joined Weight Watchers to craft and follow a 1,600- to 1,800-calorie-per-day diet, and stuck with the diet on her own for over four years. But to show how hard it's been, she now admits that she's "prouder of my weight loss than my Oscar."

Any Good News on the Horizon?

An optimistic outlook is that everything can work out if we just try hard enough and give it enough time. After all, 28 years of six successive obese US presidents, from 1884 to 1912, have been followed by more than a century without another one. Since the mid-1970s, the per-capita consumption of red meat has declined in the United States. As a result, our average blood cholesterol levels also have declined, independently of cholesterol-lowering medications and despite the fact that our overall caloric consumption and weight have increased.

Many of us also may have heard that rates of obesity recently declined slightly in preschoolers and elementary school students, presumably because of a 5 percent or so decline in their caloric intake. But more comprehensive analyses indicate that the only good news is a stabilization rather than an increase in obesity — we've actually seen no nationwide changes in childhood obesity in the last five or ten years, and obesity rates remain higher than they were back in 1999.

Wouldn't it be nice if each overweight and obese person could find a successful diet that's easy to follow? One possibility relates to how, as we discussed earlier in this chapter, mice weigh less if they can eat for only 8 hours per day rather than throughout the day. Even more impressively, when mice who previously had 24-hour access to food are restricted to 8-hour access, they actually lose weight even if they continue to consume the same number of calories. These surprising results have been replicated in fruit flies, but they've not yet been replicated in humans. Nevertheless, the possibility that we might lose weight by changing the timing rather than the amount of caloric intake may someday lead to new and more successful weight-loss diets.

Another possibility is to provide more effective dietary reminders and feedback. It's now possible to detect a person's intake of specific dietary sugars by analyzing that person's blood. Perhaps someday we'll have implantable alarms that inform and even punish us when we've eaten too much of the wrong things and reward us when we do better.

But speculation aside, the unfortunate reality is that obesity continues to increase worldwide, with no country showing declines in weight over the past three decades. In the United States, rates of diabetes doubled between 1980 and 2012 but are at least stabilizing at more than 8 percent of all adults, consistent with a similar plateauing of obesity rates in most states. We can't give up on individual dieting, but we also can't count on it to be the solution to a worldwide epidemic.

EXERCISE

Physical activities can be grouped into well-known categories according to their main physiologic effects: aerobic (or "cardio"), such as walking, running, and swimming; muscle strengthening (or "isometric"), such as weight lifting; flexibility (or stretching); and balance (for example, tai chi). Adults should perform moderate-intensity aerobic activity for at least 150 minutes per week or vigorous-intensity activities for at least 75 minutes per week — or a combination of the two, assuming that one vigorous-intensity minute is the equivalent of two moderate-intensity minutes. Examples of moderate-intensity physical activity include brisk walking at a pace of three or more miles per hour, gardening, water aerobics, doubles tennis, bicycling at a pace less than ten miles per hour, and ballroom dancing. Examples of vigorous-intensity physical activity include rapid walking, running, bicycling, lap swimming, singles tennis, heavy gardening, and hiking uphill or with a heavy backpack.

All forms of exercise can help us lose weight, but it's cardio that has by far the most benefit. Two hours of briskly walking seven miles consumes about 500–600 calories, and jogging six miles over the course of an hour burns 500–700 calories — about the same as an hour on a StairMaster. As I mentioned in chapter 2, lack of exercise is a "permissive" factor in the development of obesity — if we stay active enough, we won't continue to gain weight year after year unless we overeat substantially. Unfortunately, however, the 500–700 or so calories burned by dedicated exercisers is in the same range as a Big Mac washed down with a medium Dr Pepper, a Starbucks multigrain bagel with a mocha

Frappuccino, two Dunkin' Donuts doughnuts, or two Snickers bars. The ready availability of such tempting options is just one of the reasons why it's so hard to control our weight in the modern world.

Although increased aerobic exercise should be part of any rational weight-loss plan, exercise alone usually results in rather minimal weight loss. But exercise can be especially beneficial in helping us maintain weight loss, because we tend to have difficulty sustaining the degree of calorie reduction that achieved the weight loss in the first place. Of course, we have to be careful not to use exercise as an excuse—just because I ran three miles today doesn't mean I should reward myself with an ice cream sundae.

In addition, regular aerobic exercise reduces the risk of developing diabetes, heart disease, stroke, and blood clots in our veins. Increasing physical activity can improve outcomes in people who already have heart disease or have had a stroke, with a magnitude of benefit similar to that of blood-thinning medications. Exercise also appears to alter our epigenetics by changing the methylation of genes that regulate our fat and muscle cells. For example, it stimulates our typical white fat cells—which do little other than store fat—to change into brown fat cells, which consume energy and burn calories, thereby potentially reducing fat storage and weight.

Unfortunately, Americans' average physical activity levels have declined in the last 60 years. About 40 percent of the US population is now sedentary, and another 30 percent is less active than recommended, in large part because fewer of us exert ourselves routinely at work. If we need to go somewhere, we're also more likely to drive or ride than to walk. For example, only about 15 percent of American children now walk or bike to school, down from more than 40 percent 40 years ago.

Trends in recreational physical activity are less clear, in part because we lie about our exercise levels almost as much as we lie about what we eat and what we weigh. For example, about 60 percent of adults claim to meet US guidelines for physical activity, but assessment of their exertion using a validated accelerometer shows that at least 25 percent and perhaps as many as 80 percent of them are not meeting these guidelines. Nevertheless, we've reached a time when relatively wealthy and well-educated Americans — the types of people who can afford to join the growing number of gyms and fitness centers — get more exercise than relatively poor, uneducated Americans, who used to get exercise at work in prior generations.

And, as for many modern trends, much of the rest of the world is following our lead. At least 30 percent of the world's population doesn't get enough exercise, with the highest rates found in industrialized countries. Worldwide, the lack of physical activity is estimated to account for about 6–10 percent of the risk of diabetes and heart disease and about 9 percent of premature deaths. In the United States, the estimate is even higher, contributing to about 11 percent of deaths from all causes.

If it's hard to lose weight, it's not really much easier to increase physical activity. Part of the reason is that our interest in exercising may be built into our genes. For example, laboratory rodents, who are known for their nocturnal use of a running wheel, can be bred to love or avoid exercise. Not surprisingly, the overexercising rodents are thinner, more efficient at metabolizing fat, and better conditioned than the average, or lazy, rodents.

These differences appear to be driven by the way the hyperactive mice's brains process dopamine. Since we know that dopamine is critical to the brain's sense of pleasure and to addictive behaviors, it seems that the hyperactive mice are somehow pursuing and presumably attaining the equivalent of the so-called runner's high that joggers and marathoners experience. Exercise can make our brains feel better and can even help reduce depression, because our exercising muscles produce an enzyme that alters a depression-inducing chemical so that it no longer can get from our bloodstreams into our brains. But just like mice, some of us seem to have a greater need to exercise, or we somehow get more gratification from it, than others. One hint as to why: data suggest that the biochemical changes that drive the behaviors of hyperactive mice may be similar to those responsible for attention-deficit/hyperactivity disorder (ADHD) in humans.

Exercise recommendations must be tailored to the individual. Simple approaches include walking to work, choosing stairs rather than elevators, and setting aside just 20 or 30 minutes per day for a brisk walk or its equivalent. Occasional studies have shown benefits from an organized approach, in which people specifically list why they want to exercise, the positive effects they expect from exercise, the obstacles to exercising, and exactly how they plan to implement an exercise regimen.

Unfortunately, real-life interventions have been distressingly ineffective for improving physical activity. A review of 30 studies that attempted to increase physical activity in children found an effect that was equivalent to about four more minutes of walking or running per day. Similarly, no intervention has proved to be widely useful and applicable for increasing physical activity levels in adults. Aggressive community-wide campaigns that encourage adults to exercise can increase physical activity levels so that they burn about 35 more calories per day than before the campaign. A personal trainer can help you burn about an extra 80 calories per day.

Even intensive interventions that incorporate physicians and exercise referrals can get only about one in 12 sedentary adults to meet recommended levels of physical activity. Hard as we might try, most of us don't and won't exercise as much as we should. If we as a society are going to offset our overprotective tendencies to eat too much and clot too much, it's unlikely that we'll ever do it by increasing everyone's physical activity in an era of cars, elevators, and escalators. As we'll see in chapter 8, we'll probably have to find other ways to offset the inevitable decline in the levels of physical activity for which our genes were designed.

SALT REDUCTION

Not surprisingly, many people who are diagnosed with high blood pressure often hope to avoid medications by treating the condition "naturally." Modest reductions in salt intake—especially if combined with weight reduction, avoidance of caffeinated drinks, and increased exercise—can lower systolic blood pressure by perhaps 10 mmHg and diastolic blood pressure by perhaps 5 mmHg. However, as I've already emphasized in this chapter, few of us can lose enough weight or exercise enough to make even this much of a difference over the long haul.

How about more dramatic salt reductions? As discussed in chapter 3, we'd need to reduce daily sodium intake to 600 milligrams per day—less than half the most stringent current guidelines (1.5 grams per day, recommended by the American Heart Association) and way below the US adult average of about 3.6 grams per day—if we want to eliminate or nearly eliminate high blood pressure. And once a person has high blood pressure, it's not at all clear that such dramatic reductions in sodium intake would really be beneficial.

Although our bodies certainly don't need more than 1.5–2 grams of sodium per day, the recommended sodium intake in the twenty-first century strikes a balance between avoiding real or near dehydration (and the cascade of hormones it might trigger) on the one hand and the risk of persistent untreatable high blood pressure on the other hand. Now that medications are widely available and effective in protecting those of us in whom high salt intake causes high blood pressure, it wouldn't be surprising to find that optimal human daily salt intake is higher than it was before such treatment was available. In one recent study of elderly Americans, a daily sodium intake above 2.3 grams was associated with only a slightly higher risk of death. Another recent study reported that 3–6 grams of sodium per day was associated with the best overall survival rate. Both those recent reports contradicted most earlier scientific information, which supports a goal of no more than 2.3 grams. But in every study, blood pressures rise with progressively higher sodium intakes, so if sodium intake in the 3–6 grams per day range really provides the best possible protection against dehydration and related problems, then a lot of us will have high blood pressure — and the overall safety of this higher sodium intake will depend on our adherence to effective blood pressure screening and treatment regimens. Simply put, some of us can tolerate higher sodium intakes with few side effects — principally high blood pressure — but the rest of us, or currently at least 30 percent of Americans, would be better off with a lower sodium intake and no need to take blood pressure medication. This message isn't so different from what I said in relation to weight back in chapter 2 — a *little* extra can provide a bit of a cushion to protect against future debilitating illness and isn't so bad if it doesn't cause diabetes, poorly controlled high blood pressure, or elevated cholesterol levels.

Since reasonably achievable lifestyle changes really can't lower blood pressure very much, we might ask, "Why bother?" I would suggest several reasons. First, about half of us consume more than the American adult average of 3.6 grams of sodium per day — and 90 percent of us are above the 2.3-gram recommendation. Second, the 5–10 mmHg blood pressure reduction achievable by lifestyle changes might temporarily delay the need for an individual with high blood pressure to take medication, even though it's unlikely to be sufficient in the long term. Third, when you need medication to lower your blood pressure, you'll need a lower dose or fewer medications if you adopt a healthier lifestyle. Fourth, those of us with normal blood pressure benefit from lifestyle changes that reduce our blood pressure to levels within the normal range. Fifth, as discussed in chapter 3, a lifetime of a low-sodium diet can prevent the age-related arterial stiffness that raises our blood pressure as we get older and creates the vicious cycle that ultimately no longer responds well to salt restriction—in other words, the ideal adult salt intake probably will be lower if our arteries haven't had to adapt to high sodium levels for decades. And finally, although a modestly lower sodium intake will make only a minor difference in the blood pressure of any single individual, the sum of these individual changes across 300 million Americans would have a huge effect on the overall US population. Just a 1.2 gram per day reduction in the average American's sodium intake would collectively avoid about 50,000 strokes and about 75,000 heart attacks in the United States each year. To put these benefits into perspective, they would be about equivalent to the benefits of reducing US cigarette smoking by half.

Can we reduce our salt intake to healthier levels? Absolutely! Our craving for salt is hardwired, but it can be changed. After about 8–12 weeks on a low-salt diet, many people will lose their taste for salty foods and be satisfied with a lower salt intake. Unfortunately, renewed exposure to salty foods quickly resets the desire for more salt.

Salt reduction can also be achieved through national programs that influence our food supply, especially for processed and restaurant foods. In the United Kingdom, governmental encouragement led companies to voluntarily to reduce the salt content of processed foods bought in supermarkets by 20–30 percent, with more planned reductions on the way. In Finland, which started at a somewhat higher level of about 4.6 grams of sodium consumption per day, a variety of interventions, such as reducing the amount of salt added to prepared foods, labeling the salt content of food products, and increasing public recognition of the adverse effects of excess salt, led to a 25 percent reduction in sodium consumption. Both of these countries have already begun to see corresponding reductions in heart attacks and strokes. There's no reason we can't do the same here in the United States if we have the political will.

ADDRESSING OUR ANXIETIES AND DEPRESSION

The human species has been perpetuated by those who were most successful in having children who lived to have children of their own, not by people who were the happiest, so there's no reason to think that natural selection should result in a world full of happy people. Studies in twins, both identical and fraternal, indicate that about 40–50 percent of our happiness is genetic, but another 40 percent may be linked to intentional activities over which we have control. Happiness often equates with a sense of well-being, which is linked to feelings of autonomy, competence, and relatedness. People are generally happier if they're healthier, better educated, married, have a higher social status, have a job that they like, participate in leisure activities that they enjoy, are more religious, and have more friends. Put simply, we are happier — or at least less unhappy — if we get to do what we want to do, think and are told that we do it well, and have people who seem to give a damn that we do so.

Men and women tend to be equally happy, and happiness tends to vary very little with age. Money helps — happiness tends to increase with increasing affluence — but the relationship between the two is blunted as income rises. As a result, happiness is more influenced by our relative income within our own social comparison group than by absolute income — it's more important to "keep up with the Joneses," whatever their income may be, than to maximize absolute wealth. And not surprisingly, people who are happy tend to live longer, even after taking into account their genes and environment. Of course, no one is always happy, and all of us experience setbacks. Some of us bounce back, but others experience progressive anxiety and sadness, which can even lead to suicide.

The most successful and widely used approach to help us reduce stress and deal with serious anxieties and depressive thoughts is cognitive behavioral therapy, which is based on the belief that a change in maladaptive thinking, or even how we respond to it, can lead to changes in mood and behavior. Cognitive behavioral therapy can be implemented by teaching ourselves visualization exercises, self-motivation, distraction techniques, relaxation methods, meditation, or ways to cope by reducing negative thoughts, increasing positive feedback, and setting goals.

For people with depression, cognitive behavioral therapy is often helpful in both reducing symptoms and preventing relapses, with benefits in the 30 percent range for the former and the 50 percent range for the latter. For people with anxiety disorders, it can reduce panic symptoms, social phobia, and generalized anxiety, with benefits substantially larger than those seen for depression. For people with post-traumatic stress disorder (PTSD), cognitive behavioral therapy, either individualized or in a group, is generally recommended, although it's usually less successful than it is for anxiety or even depression. Ongoing research involves using brain imaging to predict which people are likely to have better responses to cognitive behavioral therapy.

Among Americans who commit suicide, most visited a physician within the previous months or year. Unlike heart disease—for which the risk factors are well known, measurable, and reasonably accurate—the risk of serious depression is hard to predict, and it's even harder to predict which depressed people might commit suicide. For example, physicians in the psychiatric emergency department at Massachusetts General Hospital, in Boston, were no better than chance at predicting which depressed patients would ultimately commit suicide. Recently, the US military has undertaken the Army Study to Assess Risk and Resilience in Servicemembers, a comprehensive project that has enrolled thousands of troops in the hope of developing predictors for the risk of suicide and interventions to avoid it. Suicide among active-duty army troops fell in 2013 for the first time in about a decade, but that decline was probably more related to the winding down of wars in Iraq and Afghanistan than to any specific improvements in the early recognition or treatment of at-risk individuals.

One problem with psychological interventions stems from the fact that modern therapies for anxiety and depression are often delayed or misdirected because primary care doctors' diagnoses of anxiety, depression, and PTSD are accurate only about half the time. Another big problem is that people in need don't continue to attend treatment sessions that might be helpful. Unfortunately, interventions designed to improve attendance at psychotherapy sessions are of limited benefit, resulting in at most about a 20 percent absolute increase in attendance.

Whether pursued individually or in a supportive group structure, cognitive behavioral therapy certainly can be successful in treating depression and even more successful in treating anxiety. However, a large recent experiment in England showed that cognitive behavioral therapy didn't improve overall social functioning or quality of life, so we can't expect it to be the predominant solution to our current problems of psychological disability and suicide. Whether we like it or not, medications (which I'll discuss in chapter 8) will have to be a critical part of any solution.

CAN BIG BROTHER HELP?

Good health begins at home, but we're all part of a larger environment that sets the background for what we eat, how much we exercise, and how we feel about ourselves. For example, many Americans lay the blame for childhood obesity on the shoulders of parents, but they also put some of the responsibility on the food and beverage industry and on the government. Not surprisingly, political liberals generally blame industry and government more than do conservatives.

In terms of our caloric and salt intake, the magnitude of the problem combined with the general failure of self-help programs and voluntary lifestyle interventions raise the question of whether other public health interventions will be required. What should be the role of government, international agencies, and the private sector? And since information alone is obviously inadequate for changing behavior meaningfully, what else should rational yet free societies consider?

Requiring food packages and even restaurants to provide information about the caloric content of their foods hasn't been as helpful as many had hoped. In general, people see the information but don't change what they buy, even for their children. An analysis of more than one million purchases at Starbucks before and after food labeling was mandated in New York City showed that the caloric content of the average order declined by only 15 calories! These disappointing results suggest that efforts focusing on providing information about caloric content, though theoretically useful in giving consumers the data needed for informed choices, are unlikely to have any appreciable effect on obesity.

One hope — but at this point it is just a hope — is that companies and restaurants can find healthier ways to deliver satisfaction with less sugar, fat, calories, and salt, without losing customers. For example, Burger King offered french fries that had about 40 percent less fat and 30 percent fewer calories than a similar-size portion of french fries sold by McDonald's. But after less than one year, Burger King discontinued this healthier option in two-thirds of its US and Canadian restaurants because of disappointing sales. That's why it's good to see that all three of the largest soft-drink companies (Coca-Cola, PepsiCo, and Dr Pepper Snapple) agreed to work together to reduce the US consumption of calories from sugary drinks by 20 percent by 2025.

Another approach would be to capitalize on our responsiveness to changes in price. Price increases of 25 percent or more significantly reduce the purchasing of beverages and snacks, and analogous price reductions increase purchases of fruits and vegetables. In fact, the increase in caloric intake in recent decades has been driven in part by substantial inflation-adjusted decreases in the price of calorie-dense snacks and prepared foods.

Taxes, which are a form of price increase, have also helped

reduce cigarette use, especially among low-income smokers. Similarly, gasoline taxes reduce gasoline use, with each one-cent-per-gallon increase reducing gasoline consumption by about 0.2 percent. The same principle of using taxes to change consumption could be applied to sugar-sweetened drinks, doughnuts, muffins, and Big Macs, perhaps accompanied by subsidies to encourage the purchase of healthier alternatives. For example, estimates by my Columbia colleague Claire Wang suggest that a penny-per-ounce tax on sugar-sweetened beverages could save close to $2 billion per year in health care costs while simultaneously generating more than $1 billion per year in tax revenue. But such legislation hasn't been passed, and even some public health experts debate the virtues of this approach.

Despite the notable failure of Prohibition to eliminate alcohol consumption, a variety of laws and regulations have been variably effective in changing behavior in other domains. For example, bans on smoking in the workplace and in public spaces have not only reduced smoking rates in the United States by close to 30 percent in the past two decades, they have also reduced the risk of heart attacks among nonsmokers because of a reduction in their exposure to passive smoking. We're now required to wear seat belts, which reduce the risk of death or serious injury by about 50 percent and save about 13,000 lives per year. Laws requiring motorcycle helmets, though incompletely followed, save more than 1,000 lives per year.

Can laws also change food consumption, salt intake, and even exercise habits? New York City's Board of Health banned most trans fats in the city's restaurants in 2006. In 2015, the US Food and Drug Administration (FDA) declared that partially hydrogenated oils, which are the major source of dietary trans fats, are not generally recognized as safe and must be eliminated from processed foods within three years. Even the threat of federal actions can lead to major changes in food composition and consumption. For example, food companies reduced trans fats by 85 percent in the decade before the FDA ban. Recently, a number of food companies have joined what is called the National Salt Reduction Initiative, spearheaded by former mayor Michael Bloomberg of New York City, which aims to reduce the salt in their products by 25 percent. The FDA has considered exercising its power to regulate the salt content of processed foods, but it's been hindered by the fact that salt has previously been deemed a "generally recognized as safe" substance by the agency. As a result, the FDA has set criteria for defining "low sodium" and "reduced in sodium," but it has never actually exercised its right to require a reduction in the sodium content in food. So it's not surprising that there haven't been any real reductions in the sodium content of processed and restaurant food in the United States in recent years.

Attempts to ban very large sugar-sweetened drinks in New York City were unsuccessful. Even more strikingly, Mississippi, which has one of the highest obesity rates in America, actually passed legislation *prohibiting* a ban on large drinks, despite estimates that simply capping the size of portions could lead to meaningful reductions in caloric intake. The gap between proponents, such as the New York City Board of Health, and opponents of taxing sugar-sweetened beverages or banning large servings of them exemplifies our ambivalence about restricting personal freedom. The libertarian leanings of a society that values its Bill of Rights makes it unlikely that Big Brother will be our salvation.

By comparison, consider rural China, where the average sodium intake is one-third higher than it is in the United States, and high blood pressure is even more common. China has adopted an explicit national goal to reduce salt intake to US levels as soon as possible, and a country that was able to enforce a one-child policy is likely to be aggressive in doing so.

THE SAD TRUTH

We all wish we could buck the odds and succeed in our attempts to have a healthier lifestyle, which would help us avoid the obesity, diabetes, high blood pressure, anxiety and depression, and excess clotting that are the adverse consequences of our historical survival traits. We can lose weight, increase our exercise levels, reduce our salt intake, and practice visualization exercises and meditation to shed pounds, lower our blood pressure, reduce the stress in our lives, and decrease our risks of a heart attack or stroke. We should certainly follow practical, scientifically proven recommendations for healthy living.

But the sad truth is that these good ideas and intentions aren't actually working for most of us in the real world. About 35 percent of us are obese and another one-third are overweight, often despite repeated attempts to lose weight and keep it off. Even at our best, we're stymied by a series of hormones that kick in to slow down our metabolism and stimulate us to eat more, just as they did when they tried to protect our ancestors from starving. One-third of us have high blood pressure, and lifestyle changes short of Paleolithic salt shortages — for which our salt craving was designed to compensate but now overcompensates to a degree that is harmful for our long-term survival — are unlikely to change that rate appreciably. Anxiety, depression, and suicide have plagued the human species from the beginning and aren't going away. And our highly responsive clotting tendencies, which were and still are so critical for mothers, are otherwise ill suited for the inactive lifestyles of the modern developed world.

The reality is that each and every one of these traits represents a hardwired, physiologic adaptation that helped our ancestors survive and is now built into our genes. "Mind over matter" isn't totally hopeless, and our behaviors aren't as preprogrammed as a salmon's spawning. Some of us find ways to battle our genetic tendencies and "do all the right things," and some of us can change

our behavior and come closer to our goals. But some of us can't—
not because we're weak but because our genetic traits, often rein-
forced by the environment in which we live, are just too powerful.
And enough of us, both in America and worldwide, fall into this
latter category to make behavior modification alone an inadequate
strategy.

Please don't misunderstand me. We all should try to do better.
But when we can't, we shouldn't focus on guilt and blame. Instead,
let's look forward to a new era in which we can and should use our
brains to change our physiology, just as we used them to create the
industrial age and the information age.

CHAPTER 8

Changing Our Biology

Bill Clinton was notorious for his bad eating habits until he suffered a heart attack and underwent quadruple coronary artery bypass surgery in 2004. After that proverbial wake-up call, he improved his diet, lost weight, and exercised more. Knowing that these lifestyle changes were necessary but perhaps not sufficient, he also took medication to lower his cholesterol level. But it still wasn't enough — despite his newly virtuous habits, six years later he needed to have stents placed into two of his coronary arteries to keep them open. Now he is generally following a vegan diet and has lost 30 pounds — although he admits occasionally to eating fish and eggs.

Some of us may be blessed with protective genes, a natural proclivity toward good habits, or an unusual capacity to change our behaviors. But these attributes have already proven insufficient for humanity at large. The future is worrisome if our genes can't change fast enough to keep us from getting fatter, having higher blood pressure, being more anxious and depressed, and clotting too much. And it will be even more worrisome if our attempts to address those problems by altering our habits and lifestyles continue to be as challenging as they have been for Oprah Winfrey or as inadequate as they were for Bill Clinton.

So what are our options? Although it may seem politically incorrect to advocate for more medical and even surgical interventions, we really may only have two choices. We can let our health be adversely affected by our inability to change our biology as fast as we have changed our environment, or we can harness modern science and medicine to help our bodies adapt to this new environment.

Now, let's be clear. I am not talking about eugenics, superheroes, or programmed robots. But I am advocating for ways to develop and apply modern scientific discoveries for the benefit of even those of us who are the most genetically fortunate and the most behaviorally disciplined. After all, it would be hard to find anyone who is blessed with both perfect biology and ideal habits, and even such a perfect specimen undoubtedly would have imperfect loved ones who will suffer from some of our four overprotective traits.

WHAT TO DO TODAY

Modern medical and even surgical therapies, though far from perfect, often offer remarkably effective options for the prevention and treatment of a wide variety of common conditions. When questioned, about half of American adults and about 90 percent of Americans over age 65 report having taken at least one prescription medication in the past month, and about 30 percent of American adults take two or more medications. About 90 percent of us take a prescription medication by the time we reach the age of 65. And the leading medications — statins to lower cholesterol levels and medications to lower blood pressure in people over age 60;

antidepressant medications in people between the ages of 20 and 59 — are all linked to offsetting our historical survival traits.

Of course, no one likes to take medication if it's not necessary. Across all cultures, humans have perpetually searched for natural ways to alleviate maladies and improve their health. In the United States alone, about 40 percent of us currently report using one or more alternative or complementary therapies that aren't endorsed by scientific consensus — ranging from vitamin and mineral supplements to unproved herbal and botanical substances to massage, movement, and energy therapies. Of course, scientists can be slow to give serious consideration to unconventional approaches, slow to generate consensus, and, frankly, they can sometimes be wrong. For all these reasons, it's important to evaluate potentially useful alternative therapies, preferably in careful studies in which a randomly selected half of the participants receive placebo pills or sham treatments to see if they really work.

At the other end of the spectrum are medications and devices that are approved by the US Food and Drug Administration. Although no one would claim that the FDA's decisions, which are guided by panels of independent experts, are perfect, their endorsement requires careful documentation of a treatment's effectiveness and safety.

And the FDA has made some very good decisions. For example, it was the FDA that refused to approve thalidomide for the treatment of nausea in pregnancy, despite its growing use worldwide, because of concerns about its safety. This judgment was soon validated when thalidomide was implicated as a cause of major birth deformities, especially in Germany. The FDA also banned laetrile, an extract of apricot pits, which was promoted as a cure for cancer. Careful studies documented that not only was laetrile ineffective against cancer, it also could cause cyanide poisoning!

In the movie *Dallas Buyers Club,* Matthew McConaughey won an Academy Award for portraying AIDS-infected Ron Woodroof, who smuggled not-yet-approved drugs into the United States starting in the late 1980s. But the FDA was sobered by its need to be more responsive in situations in which available treatments are inadequate, and it's subsequently found ways to fast-track drugs to combat HIV and other conditions. In carefully controlled research settings, we can also access medications that are being evaluated for potential FDA approval — after giving our informed consent.

There may be some popular, effective, but as yet unapproved therapies to prevent or treat obesity, high blood pressure, anxiety and depression, and excess clotting. But I urge that our fascination with the potential of any of these unapproved therapies not distract us from appreciating the effectiveness and acceptable safety of therapies that have been FDA-approved. And we shouldn't hope for miracles from green coffee beans, garcinia cambogia, ginkgo biloba, or even traditional vitamin supplements. Steve Jobs is the unfortunate poster child for insisting on alternative, unproved therapies and ignoring widely accepted medical advice. His nine-month delay in accepting chemotherapy and surgery for a potentially curable form of pancreatic cancer may have cost him his life.

To get a sense of where we really stand in our battle against our overprotective traits, we first must appreciate the medical and surgical treatments that are currently available and generally recommended, including both their potential and their limitations. Then we can also consider some exciting possibilities that may be on the horizon.

Overweight and Obesity

In 2013, the American Medical Association (AMA) officially declared obesity a disease. Although some skeptics rushed to criticize doctors for trying to medicalize a condition that arguably stems from a bad habit, the AMA's rationale was rather simple: to say that obesity isn't a disease because it's the result of a chosen lifestyle — eating too much and exercising too little — would be the equivalent of claiming that lung cancer isn't a disease when it's caused by a chosen lifestyle, in this case cigarette smoking. The AMA also hoped to decrease self-blame and stigmatization by recognizing that overeating is a by-product of our inherent survival traits, and that it predisposes us to serious medical conditions, such as diabetes, that require counseling, medication, and sometimes even surgery. Although I certainly have disagreed with the AMA on a number of occasions, on this point I think they got it exactly right.

Since dieting is so often unsuccessful, wouldn't it be great if we had a pill that could blunt our appetites, reduce our intestinal absorption of calories, or somehow burn calories more rapidly? Finding a truly safe and effective diet pill is the holy grail of medicine. Unfortunately, no current diet pill remotely approaches holy-grail status. Although scientists have found medications that can address many of the 20 or so molecules and hormones that regulate appetite, other hormones and molecules rev up to offset their effects. That's because our overprotective system was designed to be redundant—with many signals reinforcing one another to be sure we eat enough.

So it's not surprising that individual medications have been disappointing. Even worse, every weight-loss drug has potentially dangerous side effects. That's why anyone considering diet medications—including over-the-counter ones—should consult a doctor. And that's why doctors generally don't recommend diet drugs until someone is truly obese (indicated by a BMI of 30 or above) or has a BMI above 27 with obesity-related complications, such as diabetes.

As of this writing, only five drugs are FDA-approved in the United States for the long-term treatment of obesity—four pills and one injectable medication. Orlistat (brand name Xenical), which is available without a prescription in the United States, reduces the intestinal absorption of fats, but the disturbing loose and oily bowel movements it can cause often "outweigh" its benefits. People who take orlistat can lose 5–8 percent of their weight, but the average loss is closer to 3 percent. The second medication, sold under the brand name Qsymia, combines topiramate, which tends to make people feel full, with phentermine, which is an appetite suppressant. People who take it lose, on average, about 9 percent of their weight. The third medication (chemical name lorcaserin, branded as Belviq) activates the brain's serotonin receptors that control appetite and results in an average loss of about 3 percent of body weight. Unfortunately, both Qsymia and Belviq have a variety of side effects, ranging from heart-valve damage to psychiatric symptoms.

The fourth drug, sold under the brand name Contrave, takes a different approach. It combines bupropion, which alters the way the brain uses dopamine (and which is also effective for treating depression, as I'll discuss later), with naltrexone, which has long been used to help people reduce their craving for alcohol and opioid drugs. Contrave reduces weight by an average of about 4.5 percent, with its major side effect being nausea.

The fifth drug is liraglutide (brand name Saxenda), which suppresses appetite, makes us feel full, and can provide an average 8 percent weight loss. Like insulin, it's injected under the skin. Liraglutide is approved only for obese individuals or for overweight people who also have diabetes, high blood pressure, or high cholesterol levels. Side effects include a variety of gastrointestinal complaints and the risk of a low blood sugar level; its long-term safety is still being evaluated.

The modest success of Contrave is consistent with the idea that excess eating is linked to brain functions other than hunger. Dopamine is a key mediator of the reward system that makes our brains happier — whether through the runner's high, discussed in chapter 7, or the pleasure we get from eating. The hypothesis is that animals overeat in part because they don't get enough of a dopamine reward from normal eating. If this lack of reward is because of inadequate dopamine or an insensitivity of dopamine receptors in the brain, then it may be possible to reduce overeating by finding medications such as bupropion that increase dopamine levels or the responsiveness of dopamine receptors in the brain.

Although the precise molecular mechanisms by which naltrexone works are uncertain, the fact that it reduces craving for food underscores that some obesity appears to be triggered and sustained by habitual cravings not unlike those for alcohol and opiates. Another potential target is adiponectin, which is produced by fat cells and which sensitizes our bodies to insulin, thereby reducing the risk of diabetes. Scientists have already developed a small molecule that appears to be able to mimic the effects of adiponectin, and this line of investigation holds promise.

Since the brain signals us when to stop eating, perhaps we could accentuate those signals to get us to stop sooner. One specific part of the brain seems to be ultimately responsible for our not wanting to eat — whether it's being told that we're full, that food doesn't taste good, that a specific food brings back bad memories, or even that we're sick. Scientists already have found ways to stimulate this part of the brain in mice, and their eating can be reduced as a result. But it's too early to know whether this approach can be extended safely to humans.

Research on taste cells provides intriguing ideas for the future. Just as we can fool a sweet-sensing taste cell with saccharin, cyclamates, or aspartame, theoretically we should be able to fool it in other ways. As postulated by my colleague Charles Zuker, one of the world's leading experts on taste, and his colleagues, imagine if a sweet-sensing taste cell could be manipulated to sense light rather than sweet. We could satisfy our desire for sweetness simply by shining a flashlight on our tongues, without even bothering with artificial sweeteners! It sounds like science fiction, but it's an example of what might be possible in the future.

Some have speculated that modern society has increased our exposure to certain substances or stimuli — so-called obesogens — that are contributing to the obesity epidemic and should be avoided.

We know that some medications can induce weight gain, and a variety of chemical toxins have been proposed as culprits. But it's too early to ascribe obesity, either on an individual or population-wide basis, to noncaloric food additives, plastics, or pesticides — and it's certainly premature to suggest that eliminating them will cure obesity.

One possible environmental obesogen, however, may be central heating. How could this be? In most humans, essentially all our fat resides in white fat cells, which store semiliquid fat composed mostly of cholesterol and triglycerides. Each of us normally has an estimated 35 billion or so fat cells, which collectively can weigh an average of about 30 pounds. As we get heavier and store more calories as fat, each fat cell can enlarge about fourfold before we need to form additional fat cells — and a severely obese person can have up to four times as many fat cells as a normal-weight person.

By comparison, brown fat cells are much more metabolically active. Brown fat is especially important for hibernating mammals and newborn humans, in whom it generates heat. If we could turn more of our adult white fat cells into brown fat cells, they could help us lose weight by burning calories rather than just storing them. And it turns out that a little bit of metabolically active brown fat — several ounces compared with tens of pounds of white fat — persists into adulthood, especially in those of us who are thin.

But even if we're not thin, at least some of our apparently white fat cells are now called beige fat cells because recent experiments show that they can be turned into energy-producing fat cells if we're exposed to cold temperatures or if our immune systems are stimulated in certain ways. For example, in a carefully controlled study, young men who slept at night at a room temperature of 66 degrees Fahrenheit had slightly more brown fat than when they slept in a room kept at 75 degrees Fahrenheit. Although four weeks of sleeping at the cooler temperature didn't result in discernable weight loss, it did improve the men's response to insulin. Other studies show that we'll burn an extra 250–300 calories if we wear a cold suit at 64.4 degrees Fahrenheit for three hours or sit in a 63-degrees-Fahrenheit room for two hours.

Our Paleolithic ancestors—including the Neanderthals who (as noted in chapter 2) passed down the gene that contributes to obesity among Mexicans and Latin Americans whose ancestors crossed the Bering Strait—often had to adapt to even colder temperatures, especially at night. So part of our obesity problem may be that essentially none of us—unless we're still faced with challenges like those faced by the indigenous Siberians discussed in chapter 6—now needs to provide nearly as much of our own body heat to protect against the cold. But it's too soon to suggest that the key to preventing or fighting obesity is to turn down the thermostat or to sleep in a cave or outside under the stars.

In the future, we also may be able to manipulate our fat cells in other ways. For example, exercise stimulates muscle cells to secrete signals that can induce white fat cells to act more like calorie-burning beige or brown fat cells. The normal *FTO* gene does the same thing, and that's why mutations in it (as discussed in chapter 2) cause weight gain. Adding a chemical called beta-aminoisobutyric acid into a mouse's drinking water can stimulate its white fat cells to become brown fat cells—the mouse will lose weight and improve its blood sugar levels as well. Another chemical, called fexaramine, acts as an imaginary meal that fools the intestinal tract into thinking we're eating. The resulting release of hormones enhances the conversion of white fat cells into brown fat cells and causes mice to lose weight.

Another promising area of research is the microbiome—the population of 100 trillion or so bacteria in our bodies, especially in our intestines. We've known for some time that livestock will gain weight faster if their food is supplemented with antibiotics. Mice who are given low-dose penicillin early in life grow up with different intestinal bacteria from those who aren't given penicillin, and these bacteria seem to change their metabolisms and subsequently cause them to gain weight (transfer of their bacteria to germ-free mice also causes weight gain). Repeated exposure of children under age two to broad-spectrum antibiotics that can substantially change the intestinal microbiome has also been associated with obesity in early childhood. When the intestines of germ-free mice are inoculated with two different sets of bacteria — one from obese mice and one from skinny mice — those that receive the bacteria from fat mice get fatter than the other mice.

One possible explanation is that some bacteria may influence the release and effects of hormones that, in turn, affect hunger and how we absorb or metabolize nutrients. For example, mice who are genetically prone to developing obesity will gain less weight if their intestines are inoculated with bacteria that are designed to express a substance that modifies dietary fat and thereby suppresses appetite.

Other interesting evidence comes from studies of artificial sweeteners — saccharine (Sweet'N Low), sucralose (Splenda), and aspartame (Equal). Mice who are given these sugar substitutes don't reliably lose as much weight as you might expect, and one of the reasons may be that the sweeteners change the body's metabolism of sugar. This abnormal metabolism can be eliminated by treating the mice with antibiotics and can be transmitted to other mice by giving them the intestinal bacteria of affected mice or people.

Intriguing human data come from studies of identical twins, in whom one study found more of a specific bacteria (called *Christensenellaceae*) in thin twins than in obese twins. When human fecal material from both kinds of twins was transferred into germ-free mice, their weight gain mirrored that of the donor — but supplementation with *Christensenellaceae* could prevent the expected weight gain in the mice who received fecal material from an obese human. Ongoing research, with some early successes in mice, is looking into whether we might manipulate our bacterial diversity — by introducing certain bacteria or killing others with antibiotics — to treat or even prevent obesity.

So far, however, the potential usefulness of manipulating the intestinal microbiome in humans—other than to treat severe diarrhea caused by the overgrowth of a bacterium called *Clostridium difficile* in people who have received broad-spectrum antibiotics that also have eradicated much of the intestine's normal bacteria—is more experimental than definitive. Although the types of bacteria seem to vary depending on diet and a person's weight, it's not yet clear whether an individual's intestinal bacteria are the cause or the result. For example, some data suggest that it's the fat in our diet that overwhelmingly drives the bacterial landscape. At this point, we still have a lot to learn about the complex interactions among our intestinal bacteria, the food we eat, and the millions of cells that our intestines shed from their own internal lining each day.

Remarkably, the best current treatment for morbid obesity and obesity-related diabetes is *bariatric surgery,* which comes in three different forms: the size of the stomach can be reduced by inserting a band, part of the stomach can be removed, or the stomach can be rerouted to bypass some of the small intestines. Stomach-banding surgery—think of it as putting on a tight belt to keep the stomach from expanding to accommodate a large meal—and stomach reduction surgery work by making us get full faster. Intestinal bypass reduces the amount of calories we can absorb from the calories we eat. Sometimes people get a combined banding and bypass operation.

Not surprisingly, the most aggressive reduction of stomach size (removal, not just banding) increases average weight loss from about 5 percent or 10 percent up to about 20 percent, which is almost as much as the 25 percent average weight loss that can be obtained by the more complicated task of bypassing part of the intestines and risking problems with the absorption of key nutrients. Successful bariatric surgery also significantly improves diabetes, and intestinal bypass may be the most effective in this regard because the rerouting also changes intestinal sugar metabolism as well as the bacteria in the intestine.

Perhaps the youngest person to undergo weight-loss surgery was a two-year-old boy who weighed 72 pounds and whose BMI declined by 41 percent, from 41 to 24, in 24 months. Another dramatic response was seen in Paul Mason, who managed to put 980 pounds on a 6-foot-4-inch frame before undergoing gastric bypass surgery. He subsequently lost nearly two-thirds of his weight, down to 336 pounds and a BMI of 41, which is still obese but no longer off the charts. Governor Chris Christie of New Jersey reportedly lost about 85 pounds after bariatric surgery.

But bariatric surgery is a major procedure with its own risks, so it can be recommended only for select individuals with very severe or morbid obesity. Furthermore, many people can find ways to "graze," or eat lots of snacks, so they can maintain caloric intake without ever needing a large meal.

Potential interventional approaches to make people feel full include: balloons that are expanded in the stomach; capsules that are ingested and expand; an implantable electrical device that's hooked up to the vagus nerve in the abdomen, where it emits electrical pulses to suppress the nerve and thereby make people feel less hungry; and aspiration therapy, in which a tube is placed through the abdominal wall and into the stomach to aspirate a portion of what's been eaten during a meal. Results from these approaches have been mixed at best, and as of this writing, only two — the balloon and the electrical device — are approved in the United States for massively obese people who also have at least one obesity-related complication and have been unable to lose weight otherwise.

Even if we can't prevent or treat obesity through diet, medicines, or surgery, much of its excess risk can be eliminated by effective control of the associated increases in blood pressure and in LDL cholesterol levels (as I will soon describe). Unfortunately, no medications currently can offset the obesity-associated increased risk of diabetes. However, possibilities may be on the way. A very rare mutation in the gene known as *SLC30A8,* which codes for a protein that influences the function of the pancreatic cells that make insulin, deactivates the gene and reduces the risk of type 2 diabetes by 60–80 percent. This mutation, which occurs in perhaps one in 10,000 individuals, otherwise hasn't been associated with any other known side effects. Perhaps someday we will be able to inactivate this gene therapeutically, so we could control all three leading adverse health effects of obesity — the diabetes as well as the high blood pressure and elevated LDL cholesterol levels.

Exercise

Unlike the hyperactive mice discussed in chapter 7, humans aren't likely to increase exercise levels by being bred to do so. Furthermore, when mice who don't like to run have been given injections to mimic the brain chemistry of mice who do, the lazy mice don't run more. Perhaps a future medication might stimulate more of us to pursue the runner's high without causing us to have ADHD or other side effects. If so, the resulting increase in exercise could help offset our natural tendencies to eat too much and clot too much. It's a goal worth pursuing for our sedentary lifestyle, but currently available medical stimulants, including those that are part of some drug regimens to suppress hunger, can't yet achieve this goal safely.

Even if we can't be made to exercise more, we still might be able to mimic some of its benefits while remaining inactive. For example, the mechanism by which exercising alters a depression-inducing chemical (as described in chapter 7) can be stimulated artificially in sedentary mice. Perhaps even couch potatoes will someday be able to get many of the chemical and other benefits currently achievable only by exercising.

High Blood Pressure

Since diet and exercise have only modest benefits in people with high blood pressure, medications must be the mainstay of treatment. The first effective and generally safe medication, which became available in the 1950s, was reserpine, which blocks adrenaline and related substances. The second was hydralazine, which lowers blood pressure by dilating arteries directly. Unfortunately, both medications often caused disabling side effects and are rarely used today.

The medications that are now used to treat high blood pressure target the hormones that, as noted in chapter 3, protect us from dehydration. Diuretics, or water pills, reduce blood pressure by tricking our kidneys into eliminating salt as well as water. Beta blockers block the release of renin, a key hormone at the first step of the cascade that induces the kidneys to hold on to salt and water. Angiotensin-converting enzyme (ACE) inhibitors block the next step, and angiotensin-receptor blockers (ARBs) interfere with a later step. Calcium-channel blockers don't affect salt and water but instead lower blood pressure by directly interfering with the molecular pathways that contract the muscular layer of arteries. Perhaps not surprisingly, the major limitation of all these medications is their risk of causing low blood pressure.

With current medications, high blood pressure can be successfully controlled in the vast majority of people. Unfortunately, however, no single medicine or predictable combination of medications is uniformly successful. That's why medical experts now usually recommend small doses of two or more classes of medications — in essence trying to retune our salt and water systems at multiple levels — rather than trying to hammer it at just one step. In this way, each medication is used at a low enough dose to avoid many of its side effects, and the cocktail has good odds of addressing the cause of the high blood pressure, regardless of a patient's particular chemical cause or combination of causes.

When high blood pressure is caused by the narrowing of an artery to a kidney because of thickening of the arterial muscle, angioplasty is recommended to dilate the artery and to allow more blood to flow through. When the kidney senses more blood flow, it no longer thinks the person is dehydrated, so it reduces its secretion of hormones that hold on to unneeded salt and water. But for reasons that aren't totally understood, if the narrowing is caused by the buildup of fatty deposits, typically in patients with advanced atherosclerosis, angioplasty is unlikely to lower blood pressure, so medicines are preferred.

For the unusual cases of high blood pressure in which medications are unsuccessful, angioplasty is not helpful, and rare tumors are not the cause, a variety of procedures and devices have been proposed. The most widely touted procedure involves threading a catheter up the femoral artery in the groin and into the renal artery that supplies blood to the kidney. Then, using radio frequency, the kidney's nerves can be selectively ablated and disabled so they won't raise blood pressure. But the largest randomized trial that evaluated this procedure showed no persistent benefit from it.

The ultimate goal is to identify and target the cause of each individual's high blood pressure by interfering with the synthesis or neutralizing the effects of whatever specific hormone may be involved. As medical science advances, more and more cases of high blood pressure are being linked to specific genes that affect our sodium and water metabolism and for which such targeted therapies will likely be developed. Many of us with high blood pressure may eventually be given just the right medication in just the right dose, rather than a nonspecific cocktail to treat our high blood pressure—personalized medicine with a greater benefit and a lower risk.

Now that high blood pressure is treatable, nobody should die from uncontrolled or malignant hypertension. Since the sequence of deleterious events related to high blood pressure usually progresses over at least a decade, its serious complications can be readily avoided by routinely measuring blood pressure at all medical visits and by promptly initiating appropriate medical follow-up and treatment.

Anxiety and Depression

Although we have an unfortunate tendency to blame mental disorders on those who suffer from them, we now recognize that most psychiatric conditions — including anxiety and depression — are caused by chemical imbalances. And since cognitive and behavioral therapy, though worth trying, are far from uniformly successful, medications have become the mainstays of treatment.

The precise modes of action of various medications for depression and anxiety are incompletely understood. Most medications to treat depression are designed to increase brain levels of serotonin or norepinephrine (a relative of adrenaline). Selective serotonin reuptake inhibitors (SSRIs) and serotonin-norepinephrine reuptake inhibitors (SNRIs) work primarily by blocking the breakdown of these substances in the brain — and, it is hoped, only in the brain — so that more of their active form persists without causing a lot of side effects in other parts of the body. Most other antidepressants tend to be less selective (they often affect dopamine and adrenaline as well as serotonin and norepinephrine), which may broaden their benefit in some people — but they're also more likely to affect other organs and hence produce more side effects as a result.

How effective are these medications? About one-third of depressed people typically improve without any medication or just with a placebo. The response rate doubles to about two-thirds with SSRIs or with SNRIs, although some individuals may need to try various medications to achieve success. Unfortunately, medications usually have minimal effects for the first week and can take six or more weeks to achieve their full benefit. That's why electroconvulsive therapy (shocking the brain to cause seizures and, secondarily, change hormonal levels or chemical reactions in the brain) is sometimes required to treat severe, disabling depression. In people who have been treated with antidepressant medication for a major depression and have responded to it, continuing medications reduces the relative risk of relapse by about half, from about 40 percent to about 20 percent.

For anxiety, the most commonly prescribed medications are benzodiazepines (such as Librium and Valium), which increase the action of gamma-aminobutyric acid (GABA), a brain neurotransmitter that has a calming effect. The beta blocker propranolol is especially effective in people who have social phobias, and it can help calm people before a specific anxiety-provoking event, such as giving a speech or a performance. Buspirone, which increases serotonin in the brain but is unrelated to SSRIs or SNRIs, is another alternative. Antidepressant medications can also be useful in people who have anxiety disorders, probably because anxiety and depression may be different manifestations of similar or overlapping chemical imbalances in the brain.

For post-traumatic stress disorder (PTSD), the most commonly used medications are the SSRIs fluoxetine and sertraline and the SNRI venlafaxine. These medications not only can treat the depressive symptoms but also can reduce the risk that patients will commit violence against others.

Prophylactic medication isn't recommended during the first month after a severe traumatic event, in part because only about 10–20 percent of patients will develop PTSD and in part to emphasize that most post-traumatic emotions represent a normal response to an extraordinary situation. Once medications are started, they must usually be continued for at least 12–24 months. The discovery that calcium leakage in a particular part of the brain is implicated in PTSD raises the possibility of new pharmacological approaches to this syndrome.

Medications are clearly more effective than behavioral and

cognitive therapies, but the two can be complementary or even synergistic. The basis for such synergistic effects was very nicely demonstrated in an interesting experiment in laboratory mice. Extinction behavioral training repeatedly exposes an animal to a stimulus that induces fear, anxiety, or even PTSD by first delivering a neutral stimulus, such as a sound, in association with a noxious stimulus, until the animal associates the two; then the neutral stimulus is repeatedly delivered alone until the animal is no longer scared by it. When this treatment is used in juvenile mice, the animal's brain is sufficiently malleable that extinction training alone can be successful. In adult mice, however, a learned behavior is so well established that extinction training, though helpful, needs to be combined with an antidepressant for the adult mouse's brain to be reprogrammed. In essence, an antidepressant makes the adult brain more malleable and susceptible to behavior change, similar to a still-developing juvenile brain.

In the future, we can hope that all doctors and health care professionals get better at making earlier and more accurate diagnoses of anxiety and depression—in part by using standard, well-validated screening tools and in part by consulting more often with experts. The unfortunate reality, however, is that today's medications for anxiety and depression are little better than they were decades ago.

The good news is that some recent options are potentially promising. For example, intravenous ketamine, which heretofore has been used only as an anesthetic agent, can be a powerful and rapidly acting antidepressant medication, especially for severely depressed and suicidal patients. Unlike the commonly used current medications, ketamine acts by blocking the receptor that binds glutamate, thereby increasing its levels in the brain. Glutamate, a substance that excites the brain, represents a new target for antidepressant medications. In small studies, intravenous ketamine also has been well tolerated and effective in rapidly reducing symptoms in patients with chronic PTSD, and an intranasal form is now being developed. However, too little glutamate has also been associated with severe psychosis, and too much glutamate can kill brain neurons. That's why ketamine, though certainly promising, must be prescribed very judiciously.

New medications that are less likely to cause the sleepiness associated with benzodiazepines are being developed and tested to modulate the brain's response to GABA. Whether these medications will prove superior to benzodiazepines for treating anxiety, or at least add to current options for treating depression without causing inappropriate fearlessness, is still uncertain.

Deep brain stimulation, which uses electrodes to stimulate brain areas thought to be related to depression, generated substantial excitement because of occasional reports of remarkable responses in drug-resistant patients. But the results of more recent trials have been discouraging, and this approach remains experimental.

Even more futuristic is the possibility that the bad memories that perpetuate PTSD could be erased chemically. In mice, it's possible to use optogenetics to put a light-activated gene label on the brain cells that are involved in making new memories. These cells can then be activated by exposing them to a laser beam. Male mice conditioned to fear a location where they've been repeatedly shocked can be programmed to no longer fear it. How? The male mice were given access to female mice in another location, and this pleasant experience created good memories. When the mice were put back into scary locations and had these good memories specifically activated by a laser beam, they found the location to be pleasant!

Right now, however, optogenetics requires using a virus to insert the light-activated gene into brain cells as well as the placement of a fiber-optic cable into the skull near enough to those cells to activate them — and we're not yet ready to do that in humans. But in the future, this approach is a potential way to address a wide range of brain functions.

Unfortunately, we still don't fully understand the anatomic, cellular, biochemical, genetic, or even epigenetic causes of depression, anxiety, and PTSD. And until basic research can unlock some of these mysteries, we're unlikely to make substantial progress in preventing and treating these often debilitating conditions.

Clotting

Medications are important, if not critical, for people who have had abnormal clotting or are at clearly increased risk for it. Since nearly every heart attack is caused by a clot in a coronary artery, heart attack survivors are routinely prescribed anticlotting drugs, which are commonly termed *blood thinners*. In the coronary arteries, medications that interfere with the ability of platelets to clump and form a plug are more effective than medications that interfere with the subsequent cascade of proteins that form a larger clot. For this reason, long-term preventive treatment focuses on aspirin, which is an effective antiplatelet agent, or on newer antiplatelet prescription medications (e.g., clopidogrel and prasugrel).

Aspirin's usefulness in relieving pain was first noted in Greece around 400 BCE, when the bark of the willow tree, which contains salicylic acid, was observed to have analgesic properties. In 1763, the same type of willow bark, when dried for three months and pulverized into a powder by Edmund Stone of Oxfordshire, was also shown to reduce fever. Aspirin (acetylsalicylic acid), which was first synthesized as a safe pharmaceutical in 1899 by Felix Hoffmann at what was then called Frederich Bayer & Company, is quickly metabolized in the body to form the active salicylic acid. By 1948, the potential antiplatelet benefits of acetylsalicylic acid were first described, followed by conclusive evidence by 1968. In the 1970s, epidemiologic evidence strongly suggested that people taking aspirin had lower risks of heart attacks and strokes, and soon thereafter these benefits were confirmed in large randomized trials.

Interestingly, other nonsteroidal anti-inflammatory drugs (including such brand names as Advil, Naprosyn, Motrin, Celebrex, Voltaren, Feldene, and the like) relieve pain and reduce fever but don't share aspirin's antiplatelet effects—and they actually somewhat increase the risk of a heart attack. Even more striking is that the antiplatelet effect of aspirin is attributable to its acetyl molecule, which isn't needed for the relief of pain or fever. This phenomenon explains why other forms of salicylate in a variety of ointments for pain relief have no effect on platelets. Because salicylate actually can interfere with the antiplatelet effects of acetylsalicylic acid, low-dose aspirin, such as a daily baby aspirin, is at least as effective as higher-dose aspirin for inhibiting platelets and reducing the risk of heart attack and stroke.

In a heart attack survivor or even a patient who has angina, an antiplatelet medication can reduce the risk of a recurrent heart attack by about 25 percent. In patients like Bill Clinton, who have had one or more coronary artery stents, more powerful antiplatelet treatment, using both aspirin and a second antiplatelet drug, is commonly recommended for at least one year, and patients with a drug-eluting stent are commonly continued on two antiplatelet agents for even longer.

Aspirin is also effective for reducing the risk of a first or recurrent stroke in patients who have a narrowing of the carotid arteries that supply blood to the brain. As for heart attack survivors, the treatment is not perfect, but risk is reduced by about 25 percent.

In people who have atrial fibrillation, which is the most common heart rhythm disturbance, the fibrillating, noncontracting atrium pumps less efficiently than a normal atrium, so pools or little eddies of blood tend to remain in the atrium, where they can spontaneously form blood clots. If a blood clot breaks loose and travels to the brain, it causes a stroke. Clots in the left atrium are a result of simple stagnation, so they're much more a function of the protein cascade than of platelet plugs, and their prevention is based on medications that interfere with that cascade rather than on antiplatelet agents. The traditional treatment for such clots is a blood thinner called warfarin (brand name Coumadin), which inhibits a specific step in the cascade of clotting proteins. The dose of warfarin and its inhibitory effect must be very carefully titrated, since it can totally block the protein cascade if used in excessive doses. When it does so, patients can develop spontaneous bleeding because of an inability to maintain the basic integrity of blood vessel walls. This problem is best exemplified by the fact that warfarin is an extremely effective rodent poison—mice and rats who eat this poison spontaneously bleed into their brains and die. Fortunately, warfarin's very specific effects—it blocks the actions of vitamin K—can be reversed within hours by administering vitamin K.

Newer drugs (for example, dabigatran [Pradaxa], rivaroxaban [Xarelto], and apixaban [Eliquis]) that inhibit other proteins in the clotting protein cascade are generally safer than warfarin because they're usually less likely to cause bleeding, and they're more effective in preventing the formation or migration of clots in patients with atrial fibrillation. The one problem that has limited their use, however, is that, as of this writing, no antidote is FDA approved to reverse their anticlotting effects in the small number of patients who bleed when taking them.

For people who develop clots in their veins, warfarin is still the standard treatment, but accumulating data suggest that some of the newer inhibitors of the clotting protein cascade are reasonable alternatives. Aspirin is minimally effective, because, as with clots in the left atrium, the problem is mostly stagnation or spontaneous clot formation rather than a precipitating injury that activates platelets.

If we add up all the people in America with diseases of their coronary arteries, strokes, and atrial fibrillation, we discover that about 20 million Americans, or about 10 percent of American adults, will clearly benefit from having some type of blood thinner to modify the natural clotting system that perpetuated our species. The major problem with medications to reduce clotting is that they make us more likely to bleed. The benefit-to-risk ratio of such medications varies from very favorable (in people who have a coronary stent inserted because of a heart attack) to very unfavorable (in those of us who are about to undergo major surgery). In the future, this ratio may tilt more in favor of anticlotting medications if ongoing research can find drugs that inhibit excessive, unneeded clotting without increasing serious bleeding.

The high proportion of people who benefit from anticlotting drugs raises another question: If heart disease and stroke are the two leading causes of death, and if blood thinners such as aspirin reduce our risk of getting them, should a lot more of us be taking an aspirin or a blood thinner of some sort—perhaps except when we're about to undergo surgery? At least for now, the answer is no. Careful scientific studies have evaluated whether a daily baby aspirin is beneficial in people without evidence of coronary heart disease or strokes. What these studies show is that aspirin can reduce the risk of a subsequent heart attack, but that its side effects—bleeding, especially in the intestinal tract—offset much or sometimes more than all these benefits. For people at very high risk of a heart attack but at low risk of bleeding, aspirin is recommended. For otherwise healthy middle-aged men, taking a daily baby aspirin is a *close call*—a situation in which the evidence, despite several large studies, is insufficient to make a strong recommendation one way or the other. For those of us who have a low risk of heart disease, or for middle-aged and older individuals who have a high risk of bleeding, a daily aspirin generally is not recommended.

Based on the risks of bleeding, you would think that a medication like aspirin would be especially dangerous in our ultimate bleeding-and-clotting challenge—childbirth. In fact, aspirin generally isn't recommended in pregnant women. But a baby aspirin, taken under close medical supervision, is now recommended after the first trimester in women who are at high risk for preeclampsia, which is characterized by severe high blood pressure. What a change since the Paleolithic era!

For coronary artery disease, in which rupture of an atherosclerotic fatty plaque is the precipitating event that stimulates the formation of a clot, anything that can reduce the fatty deposits that cause atherosclerosis will also reduce the likelihood of rupture and subsequent clotting. As discussed in chapter 5, the best way to reduce fatty plaque is to reduce the blood level of LDL (bad) cholesterol. And although exercise and a careful diet can lower the LDL cholesterol level by an average of about 10–20 percent, reductions of 25 percent or more can best be achieved by taking medication, principally statins (see chapter 5). That's why aggressive use of statins is recommended for all patients with preexisting coronary artery disease or atherosclerotic narrowing of an artery that supplies blood to the brain or the legs. Statins are also widely recommended for other people with elevated LDL levels, especially if they have additional risk factors—such as high blood pressure, diabetes, cigarette smoking, being male, and being over about 50 years of age.

This general recommendation raises two questions. First, can people who have preexisting heart disease or who are at high risk for it avoid taking statins by following aggressive diets—becoming vegans, like Bill Clinton—especially if they increase their exercise? And at the other extreme, if statins are so good at reducing the risk of forming atherosclerotic plaque, why shouldn't everyone take them?

The answer to the first question—Can people at high risk avoid statins?—is almost always no. As noted in chapter 7, most of us can't follow a diet to lose weight and keep it off, let alone become vegans for life.

The second question—Should we all be on a statin?—is more difficult to answer. Statins have been shown to decrease the risk of

a heart attack even in average-risk Americans while rarely causing serious side effects for at least five to ten years. Some people
will develop muscle aches and pains, although these symptoms
occur only slightly more frequently than they do among patients
who take a placebo. At the present time, however, most experts agree
it would be premature to suggest that all American adults take a
statin for two reasons. One, we still can't be sure about long-term
side effects over many decades. Two, statins' cholesterol-reducing
benefits develop quickly, so most of their usefulness can be realized by starting them somewhat later, when people are at relatively
high risk, rather than using them and being subject to their potential side effects during decades of low risk.

Although statins are the most effective type of oral cholesterol-
lowering medications currently available, modest doses reduce
LDL levels by only about 25 percent, and even the highest doses
reduce LDL levels by only about 40 percent. Such reductions
are enough to lower our risk of heart attack significantly, but
we can already do better. Mutations that inactivate a gene called
NPC1L1 — a mutation shared by about one person in 650 — reduce
the absorption of dietary cholesterol and, as a result, the risk of
heart disease. The medication ezetimibe (brand name Zetia)
inhibits the activity of the protein made by this gene and, when
added to a statin (the combined drug is brand-named Vytorin),
further reduces the risk of heart attacks. Although these two
medications often can get the LDL cholesterol level down into the
50s, wouldn't it be nice to get it so low that we might not have to
worry about heart attacks?

Such a possibility may not be that far off. Back in 1999,
researchers Helen Hobbs, Jonathan Cohen, and Ronald Victor
embarked on the Dallas Heart Study, which enrolled about 3,500
Dallas residents, about half of whom were African American, in a
population-based study of heart disease risk factors and outcomes.
They identified people with especially high or low LDL cholesterol levels and showed that mutations in a previously underappre-

ciated gene — *PCSK9* — helped explain these outliers. In people in whom a mutation inactivated one of their *PCSK9* genes, including about 2 percent of African Americans and about 3 percent of whites in the study, LDL levels were lower — 40 percent lower in African Americans and 15 percent lower in whites, with corresponding 88 percent and 47 percent reductions in their risk of developing heart disease. But then the researchers found an African American woman who had inactivation of *both* copies of her *PCSK9* gene, and her LDL cholesterol level was 14 — below the Paleolithic range! They subsequently studied this woman in great detail to see whether this double mutation caused any side effects, and so far they have found absolutely no other abnormalities in this healthy and active mother of two.

Not surprisingly, medications have already been developed to block the activity of PCSK9. In early trials, monthly subcutaneous injections of antibodies that inhibit its effects have reduced LDL levels by about 60 percent, and LDL levels can decline to as low as 25 — close to the natural level in the Dallas woman — when such antibody treatment is combined with oral medications such as a statin and ezetimibe. The drugs have been safe so far, but it will take time to be sure that they have no long-term side effects.

Another example of the modern benefit of an unusual mutation that inactivates a gene lies in the story of apolipoprotein C3 (APOC3), which participates in the formation of triglycerides, especially after eating. Blood levels of triglycerides don't seem to be nearly as important as blood levels of LDL cholesterol for causing atherosclerotic plaques and subsequent heart attacks and strokes, and some studies show them not to be important at all. Nevertheless, as noted in chapter 5, the proteins that carry triglycerides also contain cholesterol, and they end up mostly as LDL cholesterol after they deliver the triglycerides they're transporting. A reduction of APOC3 seems to result in more rapid and efficient metabolism of absorbed triglycerides and a corresponding enhanced clearance of the cholesterol-laden protein that carries them.

It turns out that perhaps one in 150 humans has one of several mutations that inactivate one copy of the *APOC3* gene. In such people, lipid levels are reduced by about 45 percent, with a corresponding reduction in the risk of a heart attack. In mice with two mutated copies of the *APOC3* gene, cholesterol levels are even lower without any evidence of adverse effects, but we don't yet know whether inactivation of both copies would be safe and more beneficial in humans.

These unusual mutations once again show the potential benefit of future treatments that might silence a common gene that presumably helped our ancestors get the nutrition they needed but that we no longer need. Medications to lower APOC3 are already in development.

It's too early to guarantee that current or future medications to lower cholesterol levels or inhibit clotting will eliminate heart attacks and strokes. Nevertheless, medications available today have already contributed substantially to the markedly reduced risk of heart attack and stroke, and newer options will undoubtedly provide even more benefit. And when medications are inadequate, coronary artery disease still can be treated with stents and bypass surgery, as in Bill Clinton's case.

The progression of success in treating and preventing heart disease and stroke can be traced by looking at recent US presidents. In 1945, Franklin Roosevelt died of a stroke caused by untreated high blood pressure at age 63. In 1955, Dwight Eisenhower had a heart attack at the age of 65 and was hospitalized for the then-standard seven weeks. That same year, senator Lyndon Johnson had a near-fatal heart attack at age 46. Fast-forward to 2013, when George W. Bush, age 67, had a single coronary artery stent placed because of a positive stress test. And just to show how much we already take some of this for granted, no one thought twice when 65-year-old presidential candidate Mitt Romney disclosed during the 2012 election campaign that he was taking both a daily low-dose aspirin and a low-dose statin.

The Cutting Edge

One potential way to change our genes more rapidly than can be achieved by natural selection is gene therapy. In gene therapy, the DNA that codes for an important missing protein is attached to what is called a *vector,* which transports the gene into our DNA, where with any luck it's incorporated into our genome so that it produces the missing or dysfunctional protein. Gene therapy has been used experimentally to treat a variety of congenital diseases that are caused by the absence or malfunction of a single gene.

Despite occasional successes in individual patients, however, gene therapy is currently approved—and only in Europe—for just one disease: a rare mutation in the lipoprotein lipase gene that's needed to metabolize fat. The most common vectors are viruses, which tend to introduce their DNA into our cells as part of their infectious process. But our natural defense mechanisms try to resist or eliminate the infection, thereby making long-term gene transfer problematic. Nevertheless, occasional successes raise the tantalizing possibility that these hurdles may be overcome.

A second potential way to correct mutations is to alter or repair our DNA or even our RNA if the DNA that generated it remains adversely mutated. For example, it's already possible to edit and correct the gene that causes muscular dystrophy—a genetic condition characterized by muscle weakness and premature death—in the first totipotent embryonic cell of a mouse. Theoretically, this technology could also be used to edit out genes we don't need or want anymore—like the *PCSK9* gene. In humans, however, there's concern about the precision of such editing and the risk of inadvertently causing damage. As a result, scientists have called for careful oversight and regulation of this technique.

A third option would be to find a way to make specific epigenetic changes that alter gene function. Our 21,000 protein-coding genes are influenced by more than 250,000 regulators, some of which are close to them on the same chromosome. If the RNA formed by these regulators could be selectively enhanced or silenced, we could change our biology for the better. We know, for example, that a baby's food preferences can be influenced by experiences in the womb. In one experiment, babies of mothers who ate garlic capsules late in pregnancy liked garlic-laden milk, whereas other babies loathed it. The same preference can be induced for vanilla, alcohol, and mints. We don't know whether these preferences will be lifelong or only transient, but they do suggest a potential role for early nurture-induced epigenetic changes. Human trials are already under way to see if disease-causing mutant RNA can be silenced by medications that inhibit its activity.

So far, gene therapy, DNA or RNA repair, and epigenetic manipulation have focused mostly on inserting or mimicking the function of a single missing or dysfunctional gene in individuals with rare diseases that are determined by very specific mutations. It will be a far greater challenge if we ever hope to apply such methods to change nearly everyone's genome, epigenome, or complex behaviors to fight the epidemics of obesity, diabetes, high blood pressure, depression, anxiety, and heart disease.

But scientists are now shifting from trying to fix the rare deleterious mutation and are broadening their horizons to include approaches that may mimic the unusual beneficial mutation. Instead of inserting something that's missing in our genes, we may be able to silence or deactivate the genes we don't need anymore. The cases of the Dallas patient with an LDL cholesterol level of 14 and Timothy Ray Brown (the AIDS patient discussed in chapter 1) represent that paradigm. In the future, we may focus on the inactivation of one copy (in the cases of *APOC3* and *SLC30A8*) or both copies (in the cases of *PCSK9* and *CCR5,* which, also as noted in chapter 1, regulates the entry of most HIV into our cells) of a gene that nearly all the rest of us have. In other words, we may be able to improve our health as much by blocking the function of genes we have but no longer need as by trying to add new genes or functions we don't have.

Drugs to block PCSK9 may soon be FDA-approved, and drugs are already being tested to block APOC3 and SLC30A8. Maraviroc, which partially blocks CCR5, is already FDA-approved, although its effects aren't strong enough to add substantially to other anti-retroviral agents in most HIV-infected patients. But preliminary experimental data suggest that it may be possible to remove a patient's T cells, inactivate the normal *CCR5* gene in the labora-tory, and reinfuse these same T cells with the inactivated gene back into the patient — and, as a result, markedly reduce or even potentially eliminate HIV in the blood. Another exciting possibil-ity, which already has been tested in monkeys, is to stimulate the development of antibodies that block both CCR5 and the alterna-tive binding site for HIV, thereby totally preventing the spread of HIV into uninfected people or cells.

We soon may be able to activate, deactivate, or even change genes by taking advantage of a bacterial protein called Cas9, which can be linked to RNA to find a specific DNA sequence that it can then snip out, in a form of genetic microsurgery. This method — which already has been used experimentally to silence, enhance, or change specific genes in live mice and even in human cells in a dish — raises the possibility of a new approach to gene therapy. Another potential approach is to let the DNA form RNA but then to alter the way the RNA forms the subsequent protein. In the meantime, even if we can't inactivate genes or alter the RNA they make, we still have many effective medications that can block or enhance the effects of the proteins they make.

What the Future May Bring

One of the great excitements of modern biology and medicine is that we're about to embark on an era of personalized health. With the ability to sequence our entire genomes, the ability to measure a variety of biomarkers, and modern imaging techniques at our disposal, we'll increasingly know our future risks for a wide range of potential diseases. This information will facilitate the appropriate use of screening tests, help guide the selection of the best therapies, and permit an accurate estimate of our prognoses. There still will be a role for large randomized trials that determine, on average, what's best for people like you in general, but these large trials will be supplemented by much more focused research to determine whether something is good or bad for people who share your particular genetic predisposition — or even just for you as an individual.

Just think how medicine will change. Today, available medications to treat obesity leave a lot to be desired, and we may never find one medication that's effective for everyone. Several types of medications can be useful for treating high blood pressure, depression, and anxiety, but no single medication is uniformly successful. To prevent excess clotting, we have effective medications, but we would like to be able to do a better job of predicting the optimal drug and dose in each individual so as to maximize the benefit-to-risk ratio.

In the future, however, choosing a medication and recommending its dose will not be based on average data for large numbers of people or trial-and-error in an individual person but rather on specific knowledge of that individual's personal characteristics, such as his or her genes, biomarkers, and imaging results. As a result, we'll be able to get just the right medicine (and dose) or other treatments to maximize benefits while minimizing potential risks. This option, which is already becoming a reality for choosing the single best, targeted, patient-specific medication to treat many cancers, will undoubtedly become more and more common in the future.

This new era of precision medicine also represents a dramatic shift in the way we should think about health care and medicine. When health insurance was first instituted, it was intended to pay for treatments for people who were already sick. Increasingly, health insurance has been broadened to cover a variety of preventive and screening examinations and diagnostic tests. With the ability to analyze a genome, health care will now begin at or even prior to birth, long before people are at high risk or have manifest disease. Instead of *secondary prevention,* which is the prevention of recurrent or progressive disease in someone who already has it, or even *primary prevention,* which is the effort to prevent the first onset of symptoms or a disease, our efforts will shift to *primordial prevention,* which is the reversal of risk factors long before people are even at substantial risk for developing a disease.

Does precision medicine mean that, by definition, good behaviors will become unimportant because everything can be fixed by silencing or inserting a gene or by blocking or augmenting the protein it produces? Of course not. Healthy habits will always be preferable to less healthy ones.

First, a good diet has benefits for our gastrointestinal systems and livers, and obesity has detrimental effects — such as arthritis and liver problems — that go beyond diabetes and heart disease. Second, exercise has a variety of benefits on muscle mass, flexibility, endurance, a sense of well-being, and probably memory, especially as we age. Third, doing things naturally should always be preferable to relying on medications. Fourth, it seems highly unlikely that we'll all get perfect medications or genetic micro-surgery that will optimize our DNA and allow us to live risk-free.

So let me be clear. I'm not recommending or predicting a world in which sloth and gluttony should be ignored or rewarded. But we're increasingly understanding that many of our maladies are not related to individual gluttony, sloth, or mental weakness but rather are caused by genetic predispositions over which we often have some but not full control. With this reality comes the recognition that sometimes even the best of us can't simply raise our personal performance to overcome our genes. As a society, we need to become less judgmental of one another and oftentimes less critical of ourselves. We don't want to medicalize every trivial flaw, yet we also don't want to suffer because we failed to take advantage of medical treatments that could offset or reverse our genetic or behavioral limitations.

I'm also not advocating a brave new world in which we are pre-scribed unnecessary medications. But remember that about 50 percent of Americans, including about 90 percent of people over the age of 65, have taken a prescription drug within a given 30-day period. The goal is not to make us all zombies on drugs but rather to promote a judicious use of current and exciting future therapies that can help us alter our biology.

At a time when food and salt are overabundant, physical activity is required less and less, anxiety and depression are so com-mon, and we clot too much, medications will become an increasingly important way to offset the adverse effects of the mismatch between our historic survival traits and the world in which we live. We have to develop better ways not only to improve our behavior, with healthier diets and increased exercise, but also to improve our biology, with medications and procedures that either compensate for our inability to change our genes fast enough or that actually alter how our genes work.

The challenge is to use our brains, which so rapidly changed our environment and created these problems in the first place, to help get us back into sync. It won't be easy. But compared with what humans have successfully overcome for 200,000 years, the odds are on our side.

Acknowledgments

I have done my best to make *Too Much of a Good Thing* as scientifically rigorous as possible while still being understandable and, I hope, enjoyable for a broad audience. In pursuit of my first goal, I was ably assisted by Maribel Lim and Andrew Vagelos, who helped me check and double-check references, as well as by experts in the various fields covered in this book. Nevertheless, I take full responsibility for any inadvertent imperfections, some of which may be unavoidable because evolutionary biology and medical science are rapidly changing fields in which current understandings can be superseded by new knowledge. In relation to my second goal, I am grateful to Burtt Ehrlich, Peter Bernstein, and Amy Bernstein, who helped me transform my early thoughts into a coherent narrative. My children and their spouses — Jeff, Abbey, Daniel, Robyn, and Tobin — provided helpful thoughts and needed encouragement. My editor, Tracy Behar, taught me so much I didn't know about writing for a general audience.

No author works in a vacuum. As I began to refine my ideas, my search of the literature revealed a number of interesting and helpful discussions of the evolutionary basis of many modern maladies. The most important of these sources are listed in the bibliography; some of them are also explicitly cited in the text and referenced in the notes. Each of my more than 950 citations informed me and, I hope, made *Too Much of a Good Thing* a better book.

But most of all, I must thank my wife and best friend, Jill. As a genetic counselor and careful reader, she was a critical sounding board whose input was essential to both my goals. Discussions with her served as the inspiration for this book, and her encouragement, tempered by a healthy dose of targeted critiques, kept me focused.

Notes

Introduction

Leading causes of mortality: World health rankings. Health profile: United States. http://www.worldlifeexpectancy.com/country-health-profile/united-states. Accessed 7/29/14. • Hoyert DL, Xu J. *Natl Vital Stat Rep.* 2012;61:1. • Calle EE, et al. *N Engl J Med.* 2003;348:1625. • Deep vein thrombosis (DVT) / pulmonary embolism (PE)—blood clot forming in a vein. http://www.cdc.gov/ncbddd/dvt/data .html. Accessed 11/8/14. • Centers for Disease Control and Prevention (CDC). *Morb Mortal Wkly Rep.* 2005;54:628. • Jaslow R. Deaths from gastroenteritis double in U.S.: What's behind rising rates? http://www.cbsnews.com/news/deaths -from-gastroenteritis-double-in-us-whats-behind-rising-rates/. Accessed 11/8/14.

Six-year gain: Naghavi M, et al. *Lancet.* 2015;385:117. • Life expectancy. Global Health Observatory. World Health Organization. http://www.who.int/ gho/mortality_burden_disease/life_tables/situation_trends_text/en/. Accessed 8/14/14.

Chapter 1: How Our Bodies Became What They Are

Fifty-fifty chance: Appelbaum FR. In: Goldman L, Schafer AI, eds. *Goldman-Cecil Medicine;* 2016 (see bibliography).

Chronic disease: Quinn TC. In: Goldman L, Schafer AI, eds. *Goldman-Cecil Medicine;* 2016 (see bibliography).

HIV origination: Faria NR, et al. *Science.* 2014;346:56.

Homo sapiens *spread:* Stewart JR, Stringer CB. *Science.* 2012;335:1317.

Prior **Homo** *species:* Stewart JR, Stringer CB. *Science.* 2012;335:1317. • Stringer C. *Lone Survivors;* 2012 (see bibliography). • Meyer M, et al. *Science.*

2012;338:222. • Lalueza-Fox C, Gilbert MT. *Curr Biol.* 2011;21:R1002. • Wilford JN. New fossils indicate early branching of human family tree. *New York Times.* http://nyti.ms/PFqdLj. Accessed 8/8/12. • Balter M. *Science.* 2014;345:129. • Gibbons A. *Science.* 2012;337:1028. • *Science.* 2011;334:1629. • Gibbons A. *Science.* 2012;337:635.

Interbreeding: Stringer C. *Lone Survivors;* 2012 (see bibliography). • Meyer M, et al. *Science.* 2012;338:222. • Gibbons A. *Science.* 2014;343:1417. • Sankararaman S, et al. *PLoS Genet.* 2012;8:e1002947. • Palca J. Hey good lookin': Early humans dug Neanderthals. *All Things Considered.* http://www.npr.org/templates/story/story.php?storyId=126553081. Accessed 11/8/14. • Green RE, et al. *Science.* 2010;328:710. • Lachance J, et al. *Cell.* 2012;150:457. • Gibbons A. *Science.* 2014;343:471. • Sankararaman S, et al. *Nature.* 2014;507:354. • Pennisi E. *Science.* 2013;340:799.

Survival of Neanderthals, Denisovans, and humans: Stringer C. *Lone Survivors;* 2012 (see bibliography). • Meyer M, et al. *Science.* 2012;338:222. • Lalueza-Fox C, Gilbert MT. *Curr Biol.* 2011;21:R1002. • Pennisi E. *Science.* 2013;340:799. • Huerta-Sánchez E, et al. *Nature.* 2014;512:194. • Briggs AW, et al. *Science.* 2009;325:318. • Frankham R. *Annu Rev Genet.* 1995;29:305. • Palstra FP, Ruzzante DE. *Mol Ecol.* 2008;17:3428. • Waples RS. *Mol Ecol.* 2002;11:1029. • Finlayson C. *The Humans Who Went Extinct;* 2009 (see bibliography). • Mellars P, French JC. *Science.* 2011;333:623. • Higham T, et al. *Nature.* 2014;512:306.

DNA and chromosomes: Korf BR. In: Goldman L, Schafer AI, eds. *Goldman-Cecil Medicine;* 2016 (see bibliography). • *Science.* 2012;338:1528. • Pennisi E. *Science.* 2012;337:1159.

Mutations: Korf BR. In: Goldman L, Schafer AI, eds. *Goldman-Cecil Medicine;* 2016 (see bibliography). • Abecasis GR, et al. *Nature.* 2010;467:1061. • Roach JC, et al. *Science.* 2010;328:636.

Survival advantage: Relethford J. *Human Population Genetics;* 2012 (see bibliography). • Singham M. Evolution-14: How a single mutation spreads everywhere. http://blog.case.edu/singham/2007/07/25/evolution14_how_a_single_mutation_spreads_everywhere. Accessed 9/14/13.

Vitamin D: Elder CJ, Bishop NJ. *Lancet.* 2014;383:1665. • Schmid A, Walther B. *Adv Nutr.* 2013;4:453. • Garg M, et al. *Aliment Pharmacol Ther.* 2012;36:324. • Mason JB. In: Goldman L, Schafer AI, eds. *Goldman-Cecil Medicine;* 2016 (see bibliography).

Rickets: Holick MF. *Am J Clin Nutr.* 1995;61:638S. • Hess A, Unger L. *JAMA.* 1921;77:39.

Vitamin D deficiency: Afzal S, et al. *BMJ.* 2014;349:g6330. • Ginde AA, et al. *Arch Intern Med.* 2009;169:626.

Skin-lightening mutations: Holick MF. *Am J Clin Nutr.* 1995;61:638S. • Beleza S, et al. *Mol Biol Evol.* 2013;30:24. • Norton HL, et al. *Mol Biol Evol.* 2007;24:710. • Lamason RL, et al. *Science.* 2005;310:1782. • Hider JL, et al. *BMC Evol Biol.* 2013;13:150.

Vitamin D activation: Holick MF. *Am J Clin Nutr.* 1995;61:638S.

Sunburn: Thomson ML. *J Physiol.* 1951;112:31. • Thomson ML. *J Physiol.* 1951;112:22. • Costin GE, Hearing VJ. *FASEB J.* 2007;21:976.

Benefit of grandparents: Osborne DL, Hames R. *Am J Phys Anthropol.* 2014;153:1.

Folate deficiency: Jablonski NG, Chaplin G. *Proc Natl Acad Sci U S A.* 2010;107:8962. • Fukuwatari T, et al. *Biosci Biotechnol Biochem.* 2009;73:322. • Moan J, et al. *FASEB J.* 2012;26:971. • Jablonski NG, Chaplin G. *J Hum Evol.* 2000;39:57.

Skin color and latitude: Relethford J. *Human Population Genetics;* 2012 (see bibliography). • Jablonski NG, Chaplin G. *Proc Natl Acad Sci U S A.* 2010;107:8962. • Relethford JH. *Am J Phys Anthropol.* 1997;104:449.

Indian subcontinent: Reich D, et al. *Nature.* 2009;461:489.

Inuits and Arctic sunlight: Andersen S, et al. *Br J Nutr.* 2013;110:50. • Allen J. Ultraviolet radiation: how it affects life on Earth. The Earth Observatory. NASA's Earth Observing System, Project Science Office. NASA, 2001. http://earthobservatory.nasa.gov/Features/UVB/printall.php. Accessed 11/24/14.

Neanderthal skin color: Relethford J. *Human Population Genetics;* 2012 (see bibliography). • Lalueza-Fox C, et al. *Science.* 2007;318:1453.

Lactase gene: Gerbault P, et al. *Philos Trans R Soc Lond B Biol Sci.* 2011;366:863.

Lactase gene activity: Tishkoff SA, et al. *Nat Genet.* 2007;39:31.

Benefit of lactase mutations: Tishkoff SA, et al. *Nat Genet.* 2007;39:31. • Wiley AS. In: Trevathan W, et al, eds. *Evolutionary Medicine and Health;* 2008 (see bibliography). • Cooper MO, Spillman WJ. Human food from an acre of staple farm products. Farmer's Bulletin 877. Washington, DC: United States Department of Agriculture; 1917. • Cochran G, Harpending H. *The 10,000 Year Explosion : How Civilization Accelerated Human Evolution.* New York: Basic Books; 2009. • Fix AG. *Migration and Colonization in Human Microevolution.* Cambridge, UK: Cambridge University Press; 1999. • Curry A. *Nature.* 2013;500:20.

Various lactase mutations: Peeples L. Did lactose tolerance first evolve in central, rather than northern Europe? http://www.scientificamerican.com/article/lactose-toleraence. Accessed 4/2/13. • Itan Y, et al. *PLoS Comput Biol.*

2009;5:e1000491. • Check E. *Nature.* 2006;444:994. • Mattar R, et al. *Clin Exp Gastroenterol.* 2012;5:113. • Dunne J, et al. *Nature.* 2012;486:390. • Peng MS, et al. *J Hum Genet.* 2012;57:394. • Xu L, et al. *Scand J Gastroenterol.* 2010;45:168.

Niche benefit: Gerbault P, et al. *Philos Trans R Soc Lond B Biol Sci.* 2011;366:863.

Lactase mutation in Europe: Vuorisalo T, et al. *Perspect Biol Med.* 2012;55:163. • Gallego Romero I, et al. *Mol Biol Evol.* 2012;29:249.

AIDS mortality: Blankson JN, Siliciano RF. In: Goldman L, Schafer AI, eds. *Goldman-Cecil Medicine;* 2016 (see bibliography).

Long-term nonprogressor: What is the HIV+ long-term non-progressor study? National Institute of Allergy and Infectious Diseases. http://www.niaid .nih.gov/volunteer/hivlongterm/pages/default.aspx. Accessed 4/4/13. • Cohen J. *Science.* 2013;339:1134.

HIV and CCR5: Huang Y, et al. *Nat Med.* 1996;2:1240. • Elahi S, et al. *Nat Med.* 2011;17:989. • Tremblay C, et al. *Can J Infect Dis Med Microbiol.* 2013;24:202. • McGowan JP, Shah S. Understanding HIV tropism. Physicians' Research Network, 2010. http://www.prn.org/index.php/management/article/hiv_tropism _1002. Accessed 11/17/14. • Galvani AP, Novembre J. *Microbes and infection / Institut Pasteur.* 2005;7:302. • Martinson JJ, et al. *Nat Genet.* 1997;16:100. • Ionnidis JP, et al. *Ann Intern Med.* 2001;135:782.

Timothy Ray Brown: Appelbaum FR. In: Goldman L, Schafer AI, eds. *Goldman-Cecil Medicine;* 2016 (see bibliography). • Hutter G, et al. *N Engl J Med.* 2009;360:692. • Pollack A, McNeil DG Jr. In medical first, a baby with H.I.V. is deemed cured. *New York Times.* http://nyti.ms/186ERfp. Accessed 3/3/13. • Cohen J. *Science* 2013;339:1134. • Toddler "functionally cured" of HIV infection, NIH-supported investigators report. National Institute of Allergy and Infectious Diseases. http://www.niaid.nih.gov/news/newsreleases/2013/Pages/toddler functionallycured.aspx. Accessed 8/12/14. • McNeil DG Jr. Marrow transplants fail to cure two H.I.V. patients. *New York Times.* http://nyti.ms/1gM4EsS. Accessed 8/12/14.

CCR5 protection: Tremblay C, et al. *Can J Infect Dis Med Microbiol.* 2013;24:202. • Wasmuth JC, et al. *Expert Opin Drug Saf.* 2012;11:161. • Clapham PR, McKnight A. *Br Med Bull.* 2001;58:43.

CCR5 mutation history: Faure E, Royer-Carenzi M. *Infect Genet Evol.* 2008;8:864.

CCR5 mutation smallpox: Galvani AP, Slatkin M. *Proc Natl Acad Sci U S A.* 2003;100:15276. • Amsellem V, et al. *Circulation.* 2014;130:880.

Other viruses: Anthony SJ, et al. *MBio.* 2013;4. • Bausch DG. In: Goldman L, Schafer AI, eds. *Goldman-Cecil Medicine;* 2016 (see bibliography).

West Nile virus: Lim JK, et al. *J Infect Dis.* 2010;201:178.

Human transportation: Hirst KK. Animal domestication table of dates and places. http://archaeology.about.com/od/dterms/a/domestication.htm. Accessed 8/12/14. • Buchanan C. *Mixed Blessing : The Motor in Britain.* London: L. Hill; 1958.

Human population: Bloom DE. *Science.* 2011;333:562. • Wang H, et al. *Lancet.* 2014;384:957. • Naghavi M, et al. *Lancet.* 2015;385:117. • Riley JC. *Popul Dev Rev.* 2005;31:537. • The World Factbook. Country comparison: life expectancy at birth. Central Intelligence Agency. https://www.cia.gov/library/publications/the-world-factbook/rankorder/2102rank.html. Accessed 8/12/14. • Roberts L. *Science.* 2011;333:540.

Chapter 2: Hunger, Food, and the Modern Epidemics of Obesity and Diabetes

Pima diabetes: Hrdlicka A. Physiological and medical observations among the Indians of southwestern United States and northern Mexico. Washington, DC: Government Printing Office; 1908. • Joslin EP. *JAMA.* 1940;115:2033. • Parks JH, Waskow E. *Ariz Med.* 1961;18:99. • Bennett PH. *Nutr Rev.* 1999;57:S51. • Bennett PH, et al. *Lancet.* 1971;2:125.

Global diabetes rates: Global Burden of Disease Study 2013 Collaborators. *Lancet.* 2015;386:743.

Basal metabolism: Johnstone AM, et al. *Am J Clin Nutr.* 2005;82:941. • Mifflin MD, et al. *Am J Clin Nutr.* 1990;51:241. • Frankenfield DC. *Clin Nutr.* 2013;32:976. • Clugston G, et al. *Eur J Clin Nutr.* 1996;50 Suppl 1:S193. • Leonard WR. In: Stearns SC, Koella JC, eds. *Evolution in Health and Disease;* 2008 (see bibliography).

Hunter-gatherer caloric needs: Leonard WR. In: Stearns SC, Koella JC, eds. *Evolution in Health and Disease;* 2008 (see bibliography). • Balke B, Snow C. *Am J Phys Anthropol.* 1965;23:293. • Groom D. *Am Heart J.* 1971;81:304. • Hawkes K, et al. *Am Ethnol.* 1982;9:379. • Pontzer H, et al. *PLoS One.* 2012;7:e40503.

Lewis and Clark: Ambrose SE. *Undaunted Courage : Meriwether Lewis, Thomas Jefferson, and the Opening of the American West.* New York: Simon and Schuster; 1996. • Lewis M, et al. *The Journals of the Lewis and Clark Expedition.* Lincoln: University of Nebraska Press; 1983.

High-calorie burners: Calories burned during exercise, activities, sports and work. NutriStrategy. http://www.nutristrategy.com/caloriesburned.htm. Accessed 3/11/13.

Paleolithic diet: Eaton SB, Konner M. *N Engl J Med.* 1985;312:283. • Milton K. *Am J Clin Nutr.* 2000;71:665. • Cordain L, et al. *Am J Clin Nutr.* 2000;71:682.

Papua New Guinea diet: Dwyer PD, Minnegal M. *Hum Ecol.* 1991;19:187.

All-animal diet: Draper HH. *Am Anthropol.* 1977;79:309.

Hunter-gatherer cholesterol levels: Eaton SB, et al. *Am J Med.* 1988;84:739.
• Cordain L, et al. *Eur J Clin Nutr.* 2002;56 Suppl 1:S42.

Carbon isotopes: Ungar PS, Sponheimer M. *Science.* 2011;334:190.

Vitamin and mineral deficiencies: Mason JB. In: Goldman L, Schafer AI, eds. *Goldman-Cecil Medicine;* 2016 (see bibliography).

Children's developing brains: O'Dea K. *Philos Trans R Soc Lond B Biol Sci.* 1991;334:233.

Stefansson and Anderson: McClellan WS, Du Bois EF. *J Biol Chem.* 1930;87:651.

Hypothalamus and hunger: Lieberman LS. In: Trevathan W, et al, eds. *Evolutionary Medicine and Health;* 2008 (see bibliography).

Leptin: Jensen MD. In: Goldman L, Schafer AI, eds. *Goldman-Cecil Medicine;* 2016 (see bibliography).

Taste: Chaudhari N, Roper SD. *J Cell Biol.* 2010;190:285. • Stice E, et al. *Am J Clin Nutr.* 2013;98:1377.

Umami: Yarmolinsky DA, et al. *Cell.* 2009;139:234.

High salt concentration: Oka Y, et al. *Nature.* 2013;494:472.

Taste genes: Feeney E, et al. *Proc Nutr Soc.* 2011;70:135.

Saccharine taste: Behrens M, et al. *Angew Chem Int ed Engl.* 2011;50:2220.

Carbonated sodas: Chandrashekar J, et al. *Science.* 2009;326:443.

Flavor and texture: Kupferschmidt K. *Science.* 2013;340:808. • Rolls ET. *Int J Obes.* 2011;35:550.

Acquired taste: Hofmann W, et al. *Psychol Bull.* 2010;136:390. • De Houwer J, et al. *Psychol Bull.* 2001;127:853. • King BM. *Am Psychol.* 2013;68:88.

L-glutamine in breast milk: Behrens M, et al. *Angew Chem Int ed Engl.* 2011;50:2220.

Urge to eat: King BM. *Am Psychol.* 2013;68:88.

M&M experiment: Kahn BE, Wansink B. *J Consum Res.* 2004;30:519.

Traditional diets: O'Dea K. *Diabetes.* 1984;33:596. • Shintani TT, et al. *Hawaii Med J.* 2001;60:69.

Irish hunger strike: Beresford D. *Ten Men Dead : The Story of the 1981 Irish Hunger Strike.* London: Grafton; 1987.

Nonfatal starvation record: Stewart WK, Fleming LW. *Postgrad Med J.* 1973;49:203.

Agriculture and domestication of animals: Diamond JM. *Guns, Germs, and Steel : The Fates of Human Societies.* New York: W. W. Norton & Company; 1997. •

Zeder MA. *Proc Natl Acad Sci U S A.* 2008;105:11597. • Farb P, Armelagos GJ. *Consuming Passions : The Anthropology of Eating.* Boston: Houghton Mifflin; 1980.

Caloric yields of hunting and gathering and agriculture: Farb P, Armelagos GJ. *Consuming Passions : The Anthropology of Eating.* Boston: Houghton Mifflin; 1980.

Hunter-gatherer bands: Hill KR, et al. *Science.* 2011;331:1286. • Hamilton MJ, et al. *Proc Biol Sci.* 2007;274:2195. • Dyble M, et al. *Science.* 2015;348:796.

Irish potato famine: Keneally T. *Three Famines : Starvation and Politics.* New York: PublicAffairs; 2011. • Miller I. *Med Hist.* 2012;56:444.

Historical meal patterns: Olver L. Food timeline. http://www.foodtimeline .org/foodfaq7.html. Accessed 12/23/11.

Twentieth-century calories burned: Church TS, et al. *PLoS One.* 2011;6:e19657. • Ministry of Agriculture, Fisheries and Food, London (GB). National Food Survey Committee. *Household Food Consumption and Expenditure 1990 : With a Study of Trends Over the Period 1940–1990.* London: Her Majesty's Stationery Office; 1991.

Genetics of height: Lettre G. *Hum Genet.* 2011;129:465. • McQuillan R, et al. *PLoS Genet.* 2012;8:e1002655.

Paleolithic and Neolithic heights: Hermanussen M. *Hormones (Athens).* 2003;2:175.

Pigmy height: Lachance J, et al. *Cell.* 2012;150:457. • Migliano AB, et al. *Proc Natl Acad Sci U S A.* 2007;104:20216.

Protein and height: Formicola V, Giannecchini M. *J Hum Evol.* 1999;36:319.

Richard Steckel and height: Steckel RH. *Soc Sci Hist.* 2004;28:211.

Height in Holland: Max A. Dutch reach new heights. *USA Today.* http:// usatoday30.usatoday.com/news/offbeat/2006-09-16-dutch-tall_x.htm. Accessed 3/13/13.

Stone Age obese figurines: Haslam D. *Obes Rev.* 2007;8 Suppl 1:31.

Weight of American presidents: DeGregorio WA. *The Complete Book of U.S. Presidents.* 4th ed. New York: Barricade Books; 1993.

Weight of American doctors: Langfield A. Bus drivers top obese workers list; doctors tip lighter. *USA Today.* http://usat.ly/12E6tSk. Accessed 5/21/13.

Obesity and social class: Ogden CL, et al. *NCHS Data Brief.* 2010;51:1. • Eknoyan G. *Adv Chronic Kidney Dis.* 2006;13:421. • Levine JA. *Diabetes.* 2011;60:2667.

Modern caloric consumption: Swinburn BA, et al. *Lancet.* 2011;378:804. • Hall KD, et al. *PLoS One.* 2009;4:e7940.

Thomas Malthus: Malthus TR. *An Essay on the Principle of Population.* London: J. Johnson, in St. Paul's Church-yard; 1798. • Mayhew RJ. *Malthus : The Life and Legacies of an Untimely Prophet.* Cambridge, MA: The Belknap Press/Harvard University Press; 2014.

Calculating BMI: About BMI for adults. Centers for Disease Control and Prevention, 2011. http://www.cdc.gov/healthyweight/assessing/bmi/adult_bmi/. Accessed 12/28/12.

Weight gain in calories: Hill JO, et al. *Circulation.* 2012;126:126. • Hall KD, et al. *Lancet.* 2011;378:826. • Hall KD, et al. *Am J Clin Nutr.* 2012;95:989.

Weight in twins: Bouchard C, et al. *N Engl J Med.* 1990;322:1477.

Hormones in weight loss: Sumithran P, et al. *N Engl J Med.* 2011;365:1597. • Hill JO, et al. *Circulation.* 2012;126:126.

Physical activity threshold: Mayer J, et al. *Am J Clin Nutr.* 1956;4:169.

"Permissive" factor: Hill JO, et al. *Circulation.* 2012;126:126. • Hill JO, Wyatt HR. *J Appl Physiol.* 2005;99:765.

Fat cells: Hall KD, et al. *Am J Clin Nutr.* 2012;95:989.

Cost of obesity and diabetes: Adult obesity. Obesity rises among adults. Centers for Disease Control and Prevention, 2010. http://www.cdc.gov/vitalsigns/adultobesity/index.html. Accessed 2/2/15. • The cost of diabetes. American Diabetes Association, 2013. http://www.diabetes.org/advocacy/news-events/cost-of-diabetes.html. Accessed 2/2/15. • American Diabetes Association. *Diabetes Care.* 2013;36:1033.

Potential benefits of a little extra weight: Ahima RS, Lazar MA. *Science.* 2013;341:856. • Berrington de Gonzalez A, et al. *N Engl J Med.* 2010;363:2211. • Flegal KM, et al. *JAMA.* 2013;309:71. • Lavie CJ, et al. *J Am Coll Cardiol.* 2014;63:1345.

Obesity and affluence: Masters RK, et al. *Am J Public Health.* 2013;103:1895.

Diabetes statistics: Centers for Disease Control and Prevention. National diabetes fact sheet, 2011. http://www.cdc.gov/diabetes/pubs/pdf/ndfs_2011.pdf. Accessed 2/2/15. • Crandall J, Shamoon H. In: Goldman L, Schafer AI, eds. *Goldman-Cecil Medicine;* 2016 (see bibliography). • Wild S, et al. *Diabetes Care.* 2004;27:1047. • Diabetes prevalence—country rankings 2010. http://www.allcountries.org/ranks/diabetes_prevalence_country_ranks.html. Accessed 5/3/13. • Honeycutt AA, et al. *Health Care Manag Sci.* 2003;6:155.

Obesity and life expectancy: Narayan KM, et al. *JAMA.* 2003;290:1884. • Eeg-Olofsson K, et al. *Diabetologia.* 2009;52:65.

Diabetes genes: Hanson RL, et al. *Diabetes.* 2013;62:2984. • Williams RC, et al. *Diabetologia.* 2011;54:1684. • Ali O. *World J Diabetes.* 2013;4:114. • Parks BW, et al. *Cell Metab.* 2015;21:334. • Keramati AR, et al. *N Engl J Med.* 2014;370:1909.

Pima diet: DeMouy J. The Pima Indians: Pathfinders for health. National Institute of Diabetes and Digestive and Kidney Diseases. http://diabetes.niddk.nih.gov/dm/pubs/pima/pathfind/pathfind.htm. Accessed 10/15/12.

Pima diabetes: Hesse FG. *J Am Med Assoc.* 1959;170:1789. • Reid JM, et al. *Am J Clin Nutr.* 1971;24:1281. • Valencia ME, et al. *Nutr Rev.* 1999;57:S55.

Thrifty phenotype hypothesis: Hales CN, Barker DJ. *Br Med Bull.* 2001;60:5.

Thrifty genotype hypothesis: Neel JV. *Am J Hum Genet.* 1962;14:353.

FTO gene: Frayling TM, et al. *Science.* 2007;316:889. • Samaan Z, et al. *Mol Psychiatry.* 2013;18:1281.

SIM1 gene: Traurig M, et al. *Diabetes.* 2009;58:1682. • Ramachandrappa S, et al. *J Clin Invest.* 2013;123:3042. • Rong R, et al. *Diabetes.* 2009;58:478.

MRAP2 gene: Asai M, et al. *Science.* 2013;341:275.

HNF1A gene: Estrada K, et al. *JAMA.* 2014;311:2305.

SLC16A11 mutations: Williams AL, et al. *Nature.* 2014;506:97. • Mostafa N, et al. *J Comp Physiol B.* 1993;163:463. • Berg JM, et al. Triacylglycerols are highly concentrated energy stores. In: *Biochemistry.* 5th ed. New York: W. H. Freeman; 2002.

Lactose tolerance and thrifty genes: Allen JS, Cheer SM. *Curr Anthropol.* 1996;37:831.

Parental obesity: Ost A, et al. *Cell.* 2014;159:1352. • Ozanne SE. *N Engl J Med.* 2015;372:973. • Jacobson P, et al. *Am J Epidemiol.* 2007;165:101. • Oken E, et al. *Am J Obstet Gynecol.* 2007;196:322 e1.

Obesity and diabetes genes: Asai M, et al. *Science.* 2013;341:275. • Leon-Mimila P, et al. *PLoS One.* 2013;8:e70640. • Gonzalez JR, et al. *Pediatr Obes.* 2014;9:272. • Russo P, et al. *Nutr Metab Cardiovasc Dis.* 2010;20:691. • Ramachandrappa S, Farooqi IS. *J Clin Invest.* 2011;121:2080.

Modern versus hunter-gatherer diet: Cordain L, et al. *Am J Clin Nutr.* 2005;81:341.

FTO gene pre- and post-1942: Rosenquist JN, et al. *Proc Natl Acad Sci U S A.* 2015;112:354.

Estimates of caloric intake: National Research Council (US). Coordinating Committee on Evaluation of Food Consumption Surveys. Subcommittee on Criteria for Dietary Evaluation. *Nutrient Adequacy : Assessment Using Food Consumption Surveys.* Washington, DC: National Academy Press; 1986. • Centers for Disease Control and Prevention (CDC). *Morb Mortal Wkly Rep.* 2004;53:80. • Centers for Disease Control and Prevention. Intake of calories and selected nutrients for the United States population, 1999–2000. http://www.cdc.gov/nchs/data/nhanes/databriefs/calories.pdf. Accessed 5/21/13. • Swinburn B, et al. *Am J Clin Nutr.* 2009;90:1453. • Duffey KJ, Popkin BM. *PLoS Med.* 2011;8:e1001050.

Obesity in America: Finucane MM, et al. *Lancet.* 2011;377:557.

Obesity worldwide: Adult obesity—a global look at rising obesity rates. http://www.hsph.harvard.edu/obesity-prevention-source/obesity-trends/obesity-rates

-worldwide/. Accessed 5/3/13. • Ng M, et al. *Lancet.* 2014;384:766. • Cresswell JA, et al. *Lancet.* 2012;380:1325. • Stevens GA, et al. *Popul Health Metr.* 2012;10:22.

Weight in Japan: Willcox BJ, et al. *The Okinawa Program : How the World's Longest-Lived People Achieve Everlasting Health — and How You Can Too.* New York: Three Rivers Press; 2002. • Onishi N. Japan, seeking trim waists, decides to measure millions. *New York Times.* http://nyti.ms/q4yBvt. Accessed 3/13/13.

Global burden of disease: Lim SS, et al. *Lancet.* 2012;380:2224.

Chapter 3: Water, Salt, and the Modern Epidemic of High Blood Pressure

Roosevelt's medical history: Bruenn HG. *Ann Intern Med.* 1970;72:579.

Water loss: Manz F, et al. *Br J Nutr.* 2012;107:1673. • National Research Council. *Dietary Reference Intakes for Water, Potassium, Sodium, Chloride, and Sulfate.* Washington, DC: National Academies Press; 2005.

Salt loss: Lewis JL III. About body water. Merck Manual Consumer Version. 2010–2011. http://www.merckmanuals.com/home/hormonal_and_metabolic _disorders/water_balance/about_body_water.html. Accessed 2/23/15.

Running and hunting: McDougall C. *Born to Run;* 2009 (see bibliography). • Lieberman DE. *Exerc Sport Sci Rev.* 2012;40:63. • Liebenberg L. *The Art of Tracking : The Origin of Science.* Cape Town: D. Philip; 1990. • Carrier DR. *Curr Anthropol.* 1984;25:483.

Cheetahs on a treadmill: Taylor CR, Rowntree VJ. *Am J Physiol.* 1973;224:848.

Temperature in mammals: Normal rectal temperature ranges. The Merck Veterinary Manual, 2012. http://www.merckmanuals.com/vet/appendixes/ reference_guides/normal_rectal_temperature_ranges.html. Accessed 2/23/15.

Maximal sweating: Kurdak SS, et al. *Scand J Med Sci Sports.* 2010;20 Suppl 3:133.

Cooling from sweat: Properly sized room air conditioners. Energy Star. https://www.energystar.gov/index.cfm?c=roomac.pr_properly_sized. Accessed 6/12/15.

Humans versus Neanderthals: heat and calories: Steudel-Numbers KL, Tilkens MJ. *J Hum Evol.* 2004;47:95.

Thirst and the brain: Oka U, et al. *Nature.* 2015;520:349.

Thirst hormones: Slotki I, Skorecki K. In: Goldman L, Schafer AI, eds. *Goldman-Cecil Medicine;* 2016 (see bibliography).

Sweat loss and athletes: Shephard RJ. *JAMA.* 1968;205:775.

Stimulants of thirst: McKinley MJ, Johnson AK. *News Physiol Sci.* 2004;19:1.

Body sodium content: Warner GF, et al. *Circulation.* 1952;5:915.

Danger of low sodium: Slotki I, Skorecki K. In: Goldman L, Schafer AI, eds. *Goldman-Cecil Medicine;* 2016 (see bibliography) • Seifter JL. In: Goldman L, Schafer AI, eds. *Goldman-Cecil Medicine;* 2016 (see bibliography). • Thakker, RV. In: Goldman L, Schafer AI, eds. *Goldman-Cecil Medicine;* 2016 (see bibliography).

Desire for salt: Mattes RD. *Am J Clin Nutr.* 1997;65:692S.

Boy who craved salt: Wilkins L, Richter CP. *JAMA.* 1940;114:866.

Sodium loss in sweat: Taylor NA, Machado-Moreira CA. *Extrem Physiol Med.* 2013;2:4. • Taylor NA. *Compr Physiol.* 2014;4:325. • Michell AR. *Nutr Res Rev.* 1989;2:149.

Salt intake and needs: Powles J, et al. *BMJ Open.* 2013;3:e003733. • Mente A, et al. *N Engl J Med.* 2014;371:601. • Oliver WJ, et al. *Circulation.* 1975;52:146.

Kidneys' control of water: Al-Awqati Q, Barasch J. In: Goldman L, Schafer AI, eds. *Goldman-Cecil Medicine;* 2016 (see bibliography).

Story of Pheidippides: Plutarch. Were the Athenians more famous in war or in wisdom? In: Babbitt FC, trans. *Moralia.* vol. 4. Loeb Classical Library 305. Cambridge, MA: Harvard University Press; 1936.

Richard Hoyt: Yes you can! Team Hoyt. http://www.teamhoyt.com. Accessed 2/23/15. • Dick and Rick Hoyt run 32nd and last marathon. *Boston Globe.* http://www.bostonglobe.com/sports/2014/04/22/dick-and-rick-hoyt-run-marathon-their-last-duo/0802xdlCGKe5Z84VmCgMpI/story.html. Accessed 6/5/15.

Man versus horse in Wales: History made as man beats horse. BBC News. 2004. http://news.bbc.co.uk/2/hi/uk_news/wales/mid_/3801177.stm. Accessed 2/23/15.

Exercise capacity varies by temperature and humidity: Tucker R, et al. *J Physiol.* 2006;574:905. • Maughan RJ, et al. *Eur J Appl Physiol.* 2012;112:2313.

Ways to exercise longer: Lopez RM, et al. *J Strength Cond Res.* 2011;25:2944. • Burdon CA, et al. *Int J Sport Nutr Exerc Metab.* 2010;20:166. • Nybo L. *Exp Physiol.* 2012;97:333. • Mundel T, Jones DA. *Eur J Appl Physiol.* 2010;109:59.

Drinking during exercise: Noakes TD. *Ann Nutr Metab.* 2010;57 Suppl 2:9.

Gatorade and related drinks: Crowther G. Gatorade vs. Powerade: Battle of the beverage. *Northwest Runner,* 2002.

Body temperature with exercise: Kenefick RW, et al. *Sports Med.* 2007;37:312. • Gonzalez-Alonso J, et al. *J Appl Physiol.* 1999;86:1032.

Heat exhaustion: Shirreffs SM, Sawka MN. *J Sports Sci.* 2011;29 Suppl 1:S39.

Estimating exercise fluid loss: O'Neal E, et al. *Int J Sport Nutr Exerc Metab.* 2012;22:353.

Heatstroke: Sawka MN, O'Connor FG. In: Goldman L, Schafer AI, eds. *Goldman-Cecil Medicine;* 2016 (see bibliography).

Heatstroke in long-distance runners: Yankelson L, et al. *J Am Coll Cardiol.* 2014;64:463.

Rivals: Colt GH. Sibling rivalry: one long food fight. *New York Times.* http://nyti.ms/1BdPfup. Accessed 6/10/15.

Petra: Water in the Desert. American Museum of Natural History. http://www.amnh.org/exhibitions/past-exhibitions/petra/city-of-stone/water-in-the-desert. Accessed 6/10/15.

Salt through history: Kurlansky M. *Salt : A World History.* New York: Penguin Books; 2003.

Salad: Hoad TF, ed. *The Concise Oxford Dictionary of English Etymology.* Oxford: Oxford University Press; 2003.

Historical salt intakes: MacGregor G, De Wardener HE. *Salt, Diet and Health : Neptune's Poisoned Chalice — The Origins of High Blood Pressure.* Cambridge and New York: Cambridge University Press; 1998. • Beard R. The salt wars. *New York Times.* http://nyti.ms/1athYkA. Accessed 8/7/14.

Hunter-gatherer salt intake and blood pressure: Eaton SB, Konner M. *N Engl J Med.* 1985;312:283. • Intersalt. *BMJ.* 1988;297:319. • Gurven M, et al. *Hypertension.* 2012;60:25.

Blood pressure treatment guidelines: Go AS, et al. *J Am Coll Cardiol.* 2014;63:1230. • James PA, et al. *JAMA.* 2014;311:507.

Blood pressure with exercise: Kenney WL, et al. *Physiology of Sport and Exercise.* 5th ed. Champaign, IL: Human Kinetics; 2012.

Transient blood pressure increases: Victor RG. In: Goldman L, Schafer AI, eds. *Goldman-Cecil Medicine;* 2016 (see bibliography).

Blood pressure treatment in the 1940s: Moser M. *J Clin Hypertens (Greenwich).* 2006;8:15.

Atherosclerosis: Hansson GK, Hamsten A. In: Goldman L, Schafer AI, eds. *Goldman-Cecil Medicine;* 2016 (see bibliography).

Genes for blood pressure: Ehret GB, Caulfield MJ. *Eur Heart J.* 2013;34:951.

Fibromuscular dysplasia: Dubose TD Jr., Santos RM. In: Goldman L, Schafer AI, eds. *Goldman-Cecil Medicine;* 2016 (see bibliography).

High blood pressure in African American adults: High blood pressure facts. Centers for Disease Control and Prevention. http://www.cdc.gov/bloodpressure/facts.htm. Accessed 2/12/15.

Effects of excess sodium intake: Mente A, et al. *N Engl J Med.* 2014;371:601. • Cook NR, et al. *Circulation.* 2014;129:981. • Kotchen TA, et al. *N Engl J Med.* 2013;368:1229. • O'Donnell M, et al. *N Engl J Med.* 2014;371:612. • Mozaffarian D, et al. *N Engl J Med.* 2014;371:624.

Blood pressure and heart failure: O'Connor CM, Rogers J. In: Goldman L, Schafer AI, eds. *Goldman-Cecil Medicine;* 2016 (see bibliography).

Roosevelt's blood pressure history: Messerli FH. *N Engl J Med.* 1995;332:1038.

Roosevelt's doctors: Lerner BH. *Bull Hist Med.* 2007;81:386.

Blood pressure and age: Go AS, et al. *Circulation.* 2013;127:e6.

Blood pressure–related deaths: Kung HC, Xu J. *NCHS Data Brief.* 2015;193:1.

Sodium requirements and recommendations: Eckel RH, et al. *Circulation.* 2014;129:S76. • Powles J, et al. *BMJ Open.* 2013;3:e003733. • Mente A, et al. *N Engl J Med.* 2014;371:601. • World Health Organization. Guideline: Sodium intake for adults and children. http://www.who.int/nutrition/publications/guidelines/sodium_intake/en/. Accessed 8/28/14. • Winslow R. Low-salt diets may pose health risks, study finds. *Wall Street Journal.* http://on.wsj.com/14lau36. Accessed 8/14/14. • Strom BL, et al. *JAMA.* 2013;310:31.

Chapter 4: Danger, Memory, Fear, and the Modern Epidemics of Anxiety and Depression

Jason Pemberton: Barnes G. The last battle: Jason Pemberton, medically discharged from army for wounds and PTSD, killed himself and his wife. The Military Suicide Report. http://themilitarysuicidereport.wordpress.com/tag/staff-sgt-jason-pemberton/. Accessed 7/14/13. • Longa L. Murder-suicide brings to light PTSD struggles. Veterans for Common Sense. http://veteransforcommonsense.org/2012/02/07/murder-suicide-brings-to-light-ptsd-struggles/. Accessed 9/25/13. • Wilkins B. Jason Pemberton, highly-decorated Iraq War veteran with PTSD, kills wife Tiffany, himself in Daytona Beach, Florida. Moral Low Ground. http://morallowground.com/2012/02/06/jason-pemberton-highly -decorated-iraq-war-veteran-with-ptsd-kills-wife-tiffany-himself-in-daytona -beach-florida/. Accessed 7/14/13. • Mims B. Mom: Military needs to stop soldier suicides. WRAL. http://www.wral.com/lifestyles/family/story/11423095/. Accessed 7/15/13. • Gye H. Purple Heart war hero kills himself and his wife after suffering from post-traumatic stress disorder. *Daily Mail.* http://www .dailymail.co.uk/news/article-2097286/Jason-Pemberton-Purple-Heart-war -hero-kills-wife-suffering-PTSD.html. Accessed 7/14/13.

Hunter-gatherer bands: Hill KR, et al. *Science.* 2011;331:1286. • Hamilton MJ, et al. *Proc Biol Sci.* 2007;274:2195. • Dyble M, et al. *Science.* 2015;348:796.

Ötzi the Iceman: Murphy WA Jr., et al. *Radiology.* 2003;226:614. • Friend T. 'Iceman' was murdered, science sleuths say. *USA Today.* http://usatoday30.usa today.com/news/health/2003-08-11-iceman-murder_x.htm. Accessed 8/14/13.

Kennewick Man: Chatters JC. *Am Antiq.* 2000;65:291.

Myth of the gentle savage: Keeley LH. *War Before Civilization : The Myth of the Peaceful Savage.* New York: Oxford University Press; 1996.

Human violence over the ages: Pinker S. *The Better Angels of Our Nature : Why Violence Has Declined.* New York: Viking; 2011.

Hunter-gatherer violence: Hill K, Hurtado AM. *Aché Life History : The Ecology and Demography of a Foraging People.* New York: Aldine de Gruyter; 1996. • Chagnon NA. *Science.* 1988;239:985. • Hill K, et al. *J Hum Evol.* 2007;52:443.

Global homicide: Global Study on Homicide: Trends, Contexts, Data. Vienna: United Nations Office on Drugs and Crime; 2011.

Benefits to the murderer: Buss DM, Duntley JD. In: Bloom RW, Dess NK, eds. *Evolutionary Psychology and Violence;* 2003 (see bibliography). • Daly M, Wilson M. *Homicide.* New Brunswick, NJ: Transaction Books; 1988.

Homicidal thoughts: Buss DM, Duntley JD. In: Bloom RW, Dess NK, eds. *Evolutionary Psychology and Violence;* 2003 (see bibliography). • Buss DM. *The Murderer Next Door : Why the Mind Is Designed to Kill.* New York: Penguin Press; 2005.

Murders by men: Buss DM, Duntley JD. In: Bloom RW, Dess NK, eds. *Evolutionary Psychology and Violence;* 2003 (see bibliography). • Buss DM, Shackelford TK. *Clin Psychol Rev.* 1997;17:605. • Daly M, Wilson M. *Ethol Sociobiol.* 1989;10:99.

Evolutionary benefits of murder: Buss DM, Duntley JD. In: Bloom RW, Dess NK, eds. *Evolutionary Psychology and Violence;* 2003 (see bibliography). • Duntley JD, Buss DM. The plausibility of adaptations for homicide. In: Carruthers P, et al, eds. *The Structure of the Innate Mind.* New York: Oxford University Press; 2005:291. • Daly M, Wilson M. *Am Anthropol.* 1982;84:372. • Von Rueden C, et al. *Proc Biol Sci.* 2011;278:2223.

Mongol Y chromosome: Zerjal T, et al. *Am J Hum Genet.* 2003;72:717.

Intimate partner murder: Stockl H, et al. *Lancet.* 2013;382:859. • Goetz AT, et al. *Aggress Violent Behav.* 2008;13:481.

Cinderella effect: Daly M, Wilson M. *The Truth About Cinderella : A Darwinian View of Parental Love.* New Haven, Conn.: Yale University Press; 1999. • Daly M, Wilson M. *Ethol Sociobiol.* 1985;6:197. • Daly M, Wilson M. The "Cinderella effect": Elevated mistreatment of stepchildren in comparison to those living with genetic parents. http://www.cep.ucsb.edu/buller/cinderella effect facts.pdf. Accessed 7/9/13.

Zero-sum game: Daly M, Wilson M. *Crime Justice.* 1997;22:51.

Testosterone levels and risk taking: Roney JR, et al. *Proc Biol Sci.* 2010;277:57. • Trumble BC, et al. *Proc Biol Sci.* 2012;279:2907. • Ronay R, von Hippel W. *Soc Psychol Personal Sci.* 2010;1:57.

Boston teenagers: Pinker S. *The Better Angels of Our Nature : Why Violence Has Declined.* New York: Viking; 2011.

Male tennis players: Farrelly D, Nettle D. *Journal of Evolutionary Psychology.* 2007;5:141.

Murder and genetic relatives: Daly M, Wilson M. *Am Anthropol.* 1982;84:372. • Daly M, Wilson M. *Crime Justice.* 1997;22:51. • Wilbanks W. *Murder in Miami : An Analysis of Homicide Patterns and Trends in Dade County (Miami) Florida, 1917–1983.* Lanham, MD: University Press of America; 1984. • Daly M, Wilson M. *Science.* 1988;242:519.

The arms race of murder and survival: Buss DM, Duntley JD. In: Bloom RW, Dess NK, eds. *Evolutionary Psychology and Violence;* 2003 (see bibliography).

Neocortex and emotional intelligence: Goleman D. *Emotional Intelligence.* New York: Bantam Books; 1995.

Memory and Eric Kandel: Kandel ER. *J Neurosci.* 2009;29:12748.

False memories: Ramirez S, et al. *Science.* 2013;341:387.

Inability to remember: Corkin S. *Permanent Present Tense : The Unforgettable Life of the Amnesic Patient, H.M.* New York: Basic Books; 2013.

Extinguishing modern fears: Mineka S, Ohman A. *Biol Psychiatry.* 2002;52:927. • Ohman A, Mineka S. *Psychol Rev.* 2001;108:483.

Defending against acute attack: Bracha HS. *CNS Spectr.* 2004;9:679. • Nesse R. What Darwinian medicine offers psychiatry. In: Trevathan W, et al, eds. *Evolutionary Medicine.* New York: Oxford University Press; 1999.

Avoiding animal attacks: Johnson R. How to survive wild animal attacks. *Outdoor Life.* http://www.outdoorlife.com/photos/gallery/hunting/2010/02/how-survive-wild-animal-attacks. Accessed 7/8/13.

False alarms: Nesse RM. *Ann N Y Acad Sci.* 2001;935:75. • Nesse RM. *Philos Trans R Soc Lond B Biol Sci.* 2004;359:1333.

Cortisol premedication: Kaouane N, et al. *Science.* 2012;335:1510.

Calcium leaks: Liu X, et al. *Cell.* 2012;150:1055. • Shan J, et al. *J Clin Invest.* 2010;120:4388.

Telomeres: Blackburn EH, Epel ES. *Nature.* 2012;490:169.

Modern anxieties: American Psychiatric Association. *Diagnostic and Statistical Manual of Mental Disorders.* 5th ed; 2013 (see bibliography). • Kahn JP. *Angst : Origins of Anxiety and Depression.* New York: Oxford University Press; 2013. • Nesse RM. *Hum Nat.* 1990;1:261. • Keedwell P. *How Sadness Survived;* 2008 (see bibliography). • O'Donnell E. A better path to high performance. *Harvard Magazine,* 2014:11. http://harvardmagazine.com/2014/05/a-better-path-to-high-performance. Accessed 6/10/15.

Evolutionary benefit of submissiveness, sadness, and depression: Keedwell P. *How Sadness Survived;* 2008 (see bibliography).

Hunter-gatherer caloric success: Kaplan HS, et al. *Philos Trans R Soc Lond B Biol Sci.* 2009;364:3289.

GABA and depression: Challis C, et al. *J Neurosci.* 2013;33:13978.

PTSD and its effects: American Psychiatric Association. *Diagnostic and Statistical Manual of Mental Disorders.* 5th ed; 2013 (see bibliography). • Post-traumatic stress disorder (PTSD). National Institute of Mental Health, National Institutes of Health, US Department of Health and Human Services. http://www.nimh .nih.gov/health/topics/post-traumatic-stress-disorder-ptsd/index.shtml. Accessed 7/11/14. • Dincheva I, et al. *Nat Commun.* 2015;6:6395. • Friedman RA. The feel-good gene. *New York Times.* http://nyti.ms/1CJ4VuJ. Accessed 3/9/15. • Kuhn S, Gallinat J. *Biol Psychiatry.* 2013;73:70.

Fruit flies and alcohol: Shohat-Ophir G, et al. *Science.* 2012;335:1351.

Depression in twins: Nestler EJ. *Nature.* 2012;490:171.

Epigenetics and stress: Miller G. *Science.* 2010;329:24. • Amaral PP, et al. *Brief Funct Genomics.* 2013;12:254. • Kubota T, et al. *Clin Epigenetics.* 2012;4:1.

Fear and caution: Marks IM, Nesse RM. *Ethol Sociobiol.* 1994;15:247.

Bass in an aquarium: Dugatkin LA. *Behav Ecol.* 1992;3:124.

DRD4 receptor: Ding Y-C, et al. *Proc Natl Acad Sci U S A.* 2002;99:309. • Munafo MR, et al. *Biol Psychiatry.* 2008;63:197. • Thomson CJ, et al. *Scand J Med Sci Sports.* 2013;23:e108. • Grady DL, et al. *J Neurosci.* 2013;33:286. • Wu J, et al. *Mol Neurobiol.* 2012;45:605. • Berry D, et al. *Dev Psychopathol.* 2013;25:291. • Berry D, et al. *Dev Psychobiol.* 2014;56:373. • Beaver KM, et al. *Dev Psychol.* 2012;48:932. • Garcia JR, et al. *PLoS One.* 2010;5:e14162. • Matthews LJ, Butler PM. *Am J Phys Anthropol.* 2011;145:382.

Risk taking and life expectancy: Wilson M, Daly M. *BMJ.* 1997;314:1271.

Post-agriculture violence: Pinker S. *The Better Angels of Our Nature: Why Violence Has Declined.* New York: Viking; 2011. • Zerjal T, et al. *Am J Hum Genet.* 2003;72:717. • Kaplan HS, et al. *Philos Trans R Soc Lond B Biol Sci.* 2009;364:3289.

British palisades: Whittle AW. *Neolithic Europe : A Survey.* Cambridge [Cambridgeshire] and New York: Cambridge University Press; 1985. • Burgess C, et al, eds. *Enclosures and Defences in the Neolithic of Western Europe.* Oxford: British Archaeological Reports; 1988.

Homicide as a crime against a state: Elias N, et al. *The Civilizing Process : Sociogenetic and Psychogenetic Investigations.* Revised ed. Oxford and Malden, MA: Blackwell Publishers; 2000.

Change in men's faces: Gibbons A. *Science.* 2014;346:405.

Violence, evolution, and civilization: Liddle JR, et al. Evolutionary perspectives on violence, homicide, and war. In: Shackelford TK, Weekes-Shackelford VA, eds. *The Oxford Handbook of Evolutionary Perspectives on Violence, Homicide, and War.* Oxford: Oxford University Press; 2012. • Wiessner P, Pupu N. *Science.* 2012;337:1651. • Dentan RK. *The Semai : A Nonviolent People of Malaya.* New York: Holt, Rinehart; 1968.

Percent of deaths from war and genocide: Pinker S. *The Better Angels of Our Nature : Why Violence Has Declined.* New York: Viking; 2011.

Hatfields and McCoys: Waller AL. *Feud : Hatfields, McCoys, and Social Change in Appalachia, 1860–1900.* Chapel Hill: University of North Carolina Press; 1988.

Justifiable homicide in Texas: Daly M, Wilson M. *Homicide.* New Brunswick, NJ: Transaction Books; 1988.

Homicide rate in Britain: Gurr TR. *Crime Justice.* 1981;3:295.

Murder rate in the United States: United States crime rates 1960–2013. The Disaster Center. http://www.disastercenter.com/crime/uscrime.htm. Accessed 6/24/13. • Murphy SL, et al. Deaths: Final data for 2010. National Vital Statistics Reports. National Center for Health Statistics. http://www.cdc.gov/nchs/data/nvsr/nvsr61/nvsr61_04.pdf. Accessed 6/24/13.

Snakes, spiders, scorpions, and insects: Hazards to outdoor workers. Centers for Disease Control and Prevention. http://www.cdc.gov/niosh/topics/outdoor/. Accessed 7/29/14.

Auto-accident deaths: Fatality analysis reporting system (FARS) data table. National Highway Traffic Safety Administration. http://www-fars.nhtsa.dot.gov/Main/index.aspx. Accessed 7/29/14.

Who murders women: Cooper A, Smith EL. US Department of Justice, Bureau of Justice Statistics. Homicide trends in the United States, 1980–2008. http://www.bjs.gov/content/pub/pdf/htus8008.pdf. Accessed 6/10/15.

Terms for attractive women: Beneke T. *Proving Manhood : Reflections on Men and Sexism.* Berkeley: University of California Press; 1997.

Relationships between killers and killed: US Census Bureau. Statistical abstract of the United States: 2012. Section 5: Law enforcement, courts, and prisons. 131st ed. Washington, DC; 2012:199.

Prevalence of anxiety: Statistics—any anxiety disorder among adults. National Institute of Mental Health. http://www.nimh.nih.gov/health/statistics/prevalence/any-anxiety-disorder-among-adults.shtml. Accessed 6/9/15. • Kessler RC, et al. *Arch Gen Psychiatry.* 2005;62:593.

Prevalence of depression and antidepressant use: Keedwell P. *How Sadness Survived;* 2008 (see bibliography). • Major depression among adults. National Institute of Mental Health. http://www.nimh.nih.gov/health/statistics/prevalence/

major-depression-among-adults.shtml. Accessed 6/9/15. • Kessler RC, et al. *JAMA.* 2003;289:3095. • Kessler RC, et al. *Arch Gen Psychiatry.* 2005;62:593. • Mukherjee S. Post-Prozac nation. *New York Times Magazine,* April 22, 2012. http://nyti.ms/ 1AXPCvX. Accessed 6/9/14.

Depression leading to suicide: Nesse RM. *Arch Gen Psychiatry.* 2000;57:14.

Jason Pemberton: Barnes G. The last battle: Jason Pemberton, medically discharged from army for wounds and PTSD, killed himself and his wife. The Military Suicide Report. http://themilitarysuicidereport.wordpress.com/tag/ staff-sgt-jason-pemberton/. Accessed 7/14/13.

Diagnosis of PTSD: American Psychiatric Association. *Diagnostic and Statistical Manual of Mental Disorders.* 5th ed; 2013 (see bibliography). • Morris DJ. *The Evil Hours : A Biography of Post-Traumatic Stress Disorder.* Boston and New York: Eamon Dolan Books/Houghton Mifflin Harcourt; 2015.

PTSD in military personnel and dogs: Smith TC, et al. *Public Health Rep.* 2009;124:90. • Phillips CJ, et al. *BMC Psychiatry.* 2010;10:52. • Dao J. More military dogs show signs of combat stress. *New York Times.* http://nyti.ms/tyxW0F. Accessed 12/1/11.

PTSD in Nepal: Kohrt BA, et al. *Br J Psychiatry.* 2012;201:268. • Kohrt BA, et al. *JAMA.* 2008;300:691.

Military atrocities in Vietnam and Afghanistan: Goodman B. Timeline: Charlie company and the massacre at My Lai. PBS *American Experience.* http://www .pbs.org/wgbh/americanexperience/features/timeline/mylai-massacre/ 1/. Accessed 2/25/14. • Staff Sgt. Bales sentenced to life in prison for murdering 16 Afghan civilians. PBS *Newshour.* http://www.pbs.org/newshour/bb/military -july-dec13-bales_08-23/. Accessed 2/22/14.

PTSD in Vietnam veterans: Carey B. Combat stress among veterans is found to persist since Vietnam. *New York Times.* http://nyti.ms/1pFacKJ. Accessed 8/13/14.

Sati: Vijayakumar L. *Arch Suicide Res.* 2004;8:73.

Jigai: Maiese A, et al. *Am J Forensic Med Pathol.* 2014;35:8.

What leads to suicide: Suicide: Data Sources. Centers for Disease Control and Prevention. http://www.cdc.gov/violenceprevention/suicide/datasources.html. Accessed 11/3/14. • Suicide claims more lives than war, murder, and natural disasters combined. American Foundation for Suicide Prevention. http://theovernight. donordrive.com/?fuseaction=cms.page&id=1034. Accessed 11/3/14. • Risk and protective factors for suicide and suicidal behaviour: a literature review. http:// www.scotland.gov.uk/Publications/2008/11/28141444/5. Accessed 11/3/14. •

Suicide. Mental Health America. http://www.mentalhealthamerica.net/suicide. Accessed 11/3/14.

Hiwi suicide: Hill K, et al. *J Hum Evol.* 2007;52:443.

First suicide note: Thomas C. *Br Med J.* 1980;281:284.

Types of suicide: Durkheim E. *Suicide : A Study in Sociology.* Glencoe, IL: Free Press; 1951. • Shneidman ES. *The Suicidal Mind.* New York: Oxford University Press; 1996.

Alcohol, drugs, and suicide: Shneidman ES. *The Suicidal Mind.* New York: Oxford University Press; 1996.

Suicide in slaves: Piersen WD. *J Negro Hist.* 1977;62:147. • Lester D. *Suicide Life Threat Behav.* 1997;27:50.

Suicide and unemployment: Thomas K, Gunnell D. *Int J Epidemiol.* 2010;39:1464. • Reeves A, et al. *Lancet.* 2012;380:1813.

Suicide in Russia: Pridemore WA, Chamlin MB. *Addiction.* 2006;101:1719. • Perlman F, Bobak M. *Am J Public Health.* 2009;99:1818. • Pridemore WA. *Soc Forces.* 2006;85:413.

Suicide in South Korea: Age-standardized suicide rates (per 100,000). World Health Organization. http://apps.who.int/gho/data/view.main.MHSUICIDEv. Accessed 6/10/15.

Suicide in men versus women: Facts and figures. American Foundation for Suicide Prevention. http://www.afsp.org/understanding-suicide/facts-and-figures. Accessed 5/14/12.

Suicides from drug overdose: The DAWN (Drug Abuse Warning Network) Report. Emergency department visits for drug-related suicide attempts among middle-aged adults aged 45 to 64. Substance Abuse and Mental Health Services Administration. http://www.samhsa.gov/data/2K14/DAWN154/sr154-suicide-attempts-2014.htm. Accessed 8/13/14.

Gun-related homicide and suicide: Murphy SL, et al. Deaths: Final data for 2010. National Vital Statistics Reports. National Center for Health Statistics. http://www.cdc.gov/nchs/data/nvsr/nvsr61/nvsr61_04.pdf. Accessed 6/24/13.

Murder and justice in a civilized world: Rates of unsolved murder by state. The Audacious Epigone. http://anepigone.blogspot.com/2013/01/rates-of-unsolved-murder-by-state.html. Accessed 11/3/14.

Suicide rate in active-duty soldiers: Burns R. Suicides are surging among US troops. Associated Press. http://news.yahoo.com/ap-impact-suicides-surging-among-us-troops-204148055.html. Accessed 6/11/12. • Williams T. Suicides outpacing war deaths for troops. *New York Times.* http://nyti.ms/JSTRv3. Accessed

6/11/12. • Dao J, Lehren AW. Baffling rise in suicides plagues the U.S. military. *New York Times*. http://nyti.ms/YXizY6. Accessed 5/16/13.

Suicide in military veterans: Hargarten J, et al. Suicide rate for veterans far exceeds that of civilian population. Nearly one in five suicides nationally is a veteran, 49,000 took own lives between 2005 and 2011. Center for Public Integrity. http://www.publicintegrity.org/print/13292. Accessed 11/3/14. • Dao J. As suicides rise in U.S., veterans are less of total. *New York Times*. http://nyti.ms/WabGOd. Accessed 11/3/14.

Causes of suicide in soldiers: Dao J, Lehren AW. Baffling rise in suicides plagues the U.S. military. *New York Times*. http://nyti.ms/YXizY6. Accessed 5/16/13.

Suicide rates in soldiers vs. civilians: LeardMann CA, et al. *JAMA*. 2013;310:496.

Chapter 5: Bleeding, Clotting, and the Modern Epidemics of Heart Disease and Stroke

Rosie O'Donnell: O'Donnell R. My heart attack. 2012. http://rosie.com/my-heart-attack/. Accessed 10/19/12. • LeWine H. Rosie O'Donnell's heart attack a lesson for women. Harvard Health Publications. http://www.health.harvard.edu/blog/rosie-odonnells-heart-attack-a-lesson-for-women-201208225191. Accessed 10/19/12.

William Harvey: Wright T. *Circulation : William Harvey's Revolutionary Idea*. London: Chatto & Windus; 2012.

Causes of death in modern America: Murphy SL, et al. Deaths: Final data for 2010. National Vital Statistics Reports. National Center for Health Statistics. http://www.cdc.gov/nchs/data/nvsr/nvsr61/nvsr61_04.pdf. Accessed 6/24/13.

Ötzi the Iceman: Pernter P, et al. *J Archaeol Sci*. 2007;34:1784. • Gostner P, Vigl EE. *J Achaeol Sci*. 2002;29:323. • Maixner F, et al. *Cell Mol Life Sci*. 2013;70:3709.

Blood loss during delivery: Sloan NL, et al. *BJOG*. 2010;117:788. • Smith JR, Brennan BG. Postpartum hemorrhage. Medscape. http://emedicine.medscape.com/article/275038-overview. Accessed 7/29/13.

Death and uncontrolled bleeding in poorest countries: Smith JR, Brennan BG. Postpartum hemorrhage. Medscape. http://emedicine.medscape.com/article/275038-overview. Accessed 7/29/13. • Maternal mortality ratio (deaths of women per 100,000 live births). http://hdr.undp.org/en/content/maternal-mortality-ratio-deaths-100000-live-births. Accessed 6/10/15. • Leading and underlying causes of maternal mortality. UNICEF. http://www.unicef.org/wcaro/overview_2642.html. Accessed 7/29/13. • Khan KS, et al. *Lancet*. 2006;367:1066.

Oxygen and red blood cells: Bunn HF. In: Goldman L, Schafer AI, eds. *Goldman-Cecil Medicine;* 2016 (see bibliography). • Neligan P. How much oxygen is delivered to the tissues per minute? Critical Care Medicine Tutorials. http://www.ccmtutorials.com/rs/oxygen/page04.htm. Accessed 7/29/13.

Increasing oxygen delivery with exercise: Marks AR. In: Goldman L, Schafer AI, eds. *Goldman-Cecil Medicine;* 2016 (see bibliography).

Number of red blood cells: Eckert WG. *Introduction to Forensic Sciences.* 2nd ed. Boca Raton, FL: CRC Press; 1997.

Iron and hemoglobin: Bunn HF. In: Goldman L, Schafer AI, eds. *Goldman-Cecil Medicine;* 2016 (see bibliography).

Heart's response to anemia: Bunn HF. In: Goldman L, Schafer AI, eds. *Goldman-Cecil Medicine;* 2016 (see bibliography). • Understanding continuous mixed venous oxygen saturation (SvO_2) monitoring with the Edwards Swan-Ganz Oximetry TD System. Edwards Lifesciences, 2002. http://ht.edwards.com/resourcegallery/products/swanganz/pdfs/svo2edbook.pdf.

Inner lining of blood vessels: Schafer AI. In: Goldman L, Schafer AI, eds. *Goldman-Cecil Medicine;* 2016 (see bibliography).

Platelet counts with anemia: Kadikoylu G, et al. *J Natl Med Assoc.* 2006;98:398.

Clotting proteins: Dashty M, et al. *Sci Rep.* 2012;2:787.

Placental circulation: Wang Y, Zhao S. Vascular biology of the placenta: Placental blood circulation. Morgan & Claypool Life Sciences, 2010. Retrieved from http://www.ncbi.nlm.nih.gov/books/NBK53254/. Accessed 10/20/13.

Circulation to the arm and leg: Ganz V, et al. *Circulation.* 1964;30:86.

Multivitamins and iron in pregnancy: Haider BA, et al. *BMJ.* 2013;346:f3443.

High blood pressure in pregnancy: Abalos E, et al. *BJOG.* 2014;121 Suppl 1:14. • Abalos E, et al. *Eur J Obstet Gynecol Reprod Biol.* 2013;170:1. • Osungbade KO, Ige OK. *J Pregnancy.* 2011;2011:481095. • Kassebaum NJ, et al. *Lancet.* 2014;384:980.

Blood clots in pregnancy: Reducing the risk of thrombosis and embolism during pregnancy and the puerperium. 3rd ed. Royal College of Obstetricians and Gynaecologists. https://www.rcog.org.uk/globalassets/documents/guidelines/gtg-37a.pdf. Accessed 6/10/15. • Prisco D, et al. *Haematologica Reports.* 2005;1:1. • Hellgren M. *Semin Thromb Hemost.* 2003;29:125. • Simpson EL, et al. *BJOG.* 2001;108:56. • Ros HS, et al. *Am J Obstet Gynecol.* 2002;186:198. • Ray JG, Chan WS. *Obstet Gynecol Surv.* 1999;54:265.

Post-delivery hemorrhage in sub-Saharan Africa in 1990: Smith JR, Brennan BG. Postpartum hemorrhage. Medscape. http://emedicine.medscape.com/article/275038-overview. Accessed 7/29/13. • Maternal mortality ratio (deaths of women per 100,000 live births). http://hdr.undp.org/en/content/

maternal-mortality-ratio-deaths-100000-live-births. Accessed 6/10/15. • Leading and underlying causes of maternal mortality. UNICEF. http://www.unicef .org/wcaro/overview_2642.html. Accessed 7/29/13. • Khan KS, et al. *Lancet.* 2006;367:1066.

Genetics of postpartum bleeding: Oberg AS, et al. *BMJ.* 2014;349:g4984.

Factor V Leiden: van Mens TE, et al. *Thromb Haemost.* 2013;110:23. • Appleby RD, Olds RJ. *Pathology.* 1997;29:341. • Kjellberg U, et al. *Am J Obstet Gynecol.* 2010;203:469 e1.

Hemochromatosis: Milman N, Pedersen P. *Clin Genet.* 2003;64:36. • Symonette CJ, Adams PC. *Can J Gastroenterol.* 2011;25:324. • Neghina AM, Anghel A. *Ann Epidemiol.* 2011;21:1. • Datz C, et al. *Clin Chem.* 1998;44:2429. • Raddatz D, et al. *Z Gastroenterol.* 2003;41:1069. • Bacon BR. In: Goldman L, Schafer AI, eds. *Goldman-Cecil Medicine;* 2016 (see bibliography). • Steinberg KK, et al. *JAMA.* 2001;285:2216. • Phatak PD, et al. *Ann Intern Med.* 1998;129:954. • Allen KJ, et al. *N Engl J Med.* 2008;358:221. • Barton JC, Acton RT. *Genet Med.* 2001;3:294.

Iron and infections: Khan FA, et al. *Int J Infect Dis.* 2007;11:482.

Malaria and other causes of death: Estimates for 2000–2012. Cause-specific mortality. World Bank regions, GHE_DthWBR_2000_2012.xls. World Health Organization. http://www.who.int/healthinfo/global_burden_disease/ estimates/en/index1.html. Accessed 8/4/14.

Talmudic tradition: Pemberton SG. *The Bleeding Disease : Hemophilia and the Unintended Consequences of Medical Progress.* Baltimore: Johns Hopkins University Press; 2011.

Early reports of hemophilia: Pemberton SG. *The Bleeding Disease : Hemophilia and the Unintended Consequences of Medical Progress.* Baltimore: Johns Hopkins University Press; 2011. • Otto JC. *Medical Repository.* 1803;6:1.

Blood clots on plane trips: Schwarz T, et al. *Arch Intern Med.* 2003;163:2759. • Scurr JH, et al. *Lancet.* 2001;357:1485.

Varicose veins: Ginsberg J. In: Goldman L, Schafer AI, eds. *Goldman-Cecil Medicine;* 2016 (see bibliography).

Pulmonary embolism: Weitz JI. In: Goldman L, Schafer AI, eds. *Goldman-Cecil Medicine;* 2016 (see bibliography).

Healthful polyunsaturated fats: Eaton SB. *Lipids.* 1992;27:814.

Fats in modern diets: Hansson GK, Hamsten A. In: Goldman L, Schafer AI, eds. *Goldman-Cecil Medicine;* 2016 (see bibliography). • Scientific Report of the 2015 Dietary Guidelines Advisory Committee. Retrieved from http://www .health.gov/dietaryguidelines/2015-scientific-report/. Accessed 6/16/15.

How fats are carried by lipoproteins: Semenkovich CF. In: Goldman L, Schafer AI, eds. *Goldman-Cecil Medicine;* 2016 (see bibliography). • Freeman MW, Junge C. *The Harvard Medical School Guide to Lowering Your Cholesterol.* New York: McGraw-Hill; 2005.

How statins and other medications work: Sirtori CR. *Pharmacol Res.* 1990;22:555. • Cameron J, et al. *FEBS J.* 2008;275:4121.

Fissuring and rupture in a coronary artery: Libby P. *N Engl J Med.* 2013;368:2004.

Intravenous fluids: Barsoum N, Kleeman C. *Am J Nephrol.* 2002;22:284. • Millam D. *J Intraven Nurs.* 1996;19:5.

Early history of blood transfusions: Myhre BA. *Transfusion.* 1990;30:358. • History of blood transfusion. American Red Cross. http://www.redcrossblood .org/print/learn-about-blood/history-blood-transfusion. Accessed 7/29/13. • Learoyd P. A short history of blood transfusion. National Blood Service. http:// www.sld.cu/galerias/pdf/sitios/anestesiologia/history_of_transfusion.pdf. Accessed 7/29/13.

First hospital blood bank: Millam D. *J Intraven Nurs.* 1996;19:5.

Early success of the Red Cross: History of blood transfusion. American Red Cross. http://www.redcrossblood.org/print/learn-about-blood/history-blood -transfusion. Accessed 7/29/13.

Early history of venous thrombosis: Galanaud JP, et al. *J Thromb Haemost.* 2013;11:402. • Mannucci PM, Poller L. *Br J Haematol.* 2001;114:258. • Anning S. *Med Hist.* 1957;1:28.

First description of pulmonary embolism: Smith TP. *AJR Am J Roentgenol.* 2000;174:1489. • Laennec R. *De l'auscultation médiate ou traité du diagnostic des maladies des poumons et du coeur.* Paris: Brosson et Chaudé; 1819.

First description of coronary artery disease: Proudfit WL. *Br Heart J.* 1983;50:209.

Early autopsy evidence of heart attack: King LS, Meehan MC. *Am J Pathol.* 1973;73:514.

First antemortem diagnosis of heart attack: Lie JT. *Am J Cardiol.* 1978;42:849.

Heart attack data through 1940: Morris JN. *Lancet.* 1951;1:1.

Paul Dudley White's experience: White PD. *Heart Disease.* New York: Macmillan; 1944.

Arterial calcifications: Nieto FJ. Historical reflections on the inflammatory aspects of atherosclerosis. In: Wick G, Grundtman C, eds. *Inflammation and Atherosclerosis.* Vienna and New York: Springer; 2012:1. • Thompson RC, et al. *Lancet.* 2013;381:1211. • Keller A, et al. *Nat Commun.* 2012;3:698.

Fissure, rupture, and heart attacks: Anderson JL. In: Goldman L, Schafer AI, eds. *Goldman-Cecil Medicine;* 2016 (see bibliography).

Atherosclerotic plaques in young Korean soldiers: Enos WF, et al. *JAMA.* 1953;152:1090.

Historical figures with strokes: Moog FP, Karenberg A. *J Med Biogr.* 2004;12:43.

Apoplexy from bleeding and clotting: Storey CE. Apoplexy: Changing concepts in the eighteenth century. In: Whitaker HA, et al, eds. *Brain, Mind and Medicine : Essays in Eighteenth-Century Neuroscience.* New York: Springer; 2007. • Wepfer JJ. Historiae apoplecticorum. In: Major RH, ed. *Classic Descriptions of Disease : With Biographical Sketches of the Authors.* 3rd ed. Springfield, IL: Charles C. Thomas; 1978. • Lidell JA. *A Treatise on Apoplexy, Cerebral Hemorrhage, Cerebral Embolism, Cerebral Gout, Cerebral Rheumatism, and Epidemic Cerebro-spinal Meningitis.* New York: Wm. Wood & Company; 1873.

Injuries in sub-Saharan Africa: Estimates for 2000–2012. Cause-specific mortality. World Bank regions, GHE_DthWBR_2000_2012.xls. World Health Organization. http://www.who.int/healthinfo/global_burden_disease/estimates/en/index1.html. Accessed 8/4/14.

Unintentional injuries worldwide: Norton R, Kobusingye O. *N Engl J Med.* 2013;368:1723.

Ronald Reagan's survival after being shot: Beahrs OH. *J Am Coll Surg.* 1994;178:86. • Carson JL, et al. *JAMA.* 2013;309:83.

Declining death rate from medically treated injuries: Harris AR, et al. *Homicide stud.* 2002;6:128.

Declining rate of postpartum hemorrhage: Maternal mortality ratio (deaths of women per 100,000 live births). http://hdr.undp.org/en/content/maternal-mortality-ratio-deaths-100000-live-births. Accessed 6/10/15. • Anderson JM, Etches D. *Am Fam Physician.* 2007;75:875. • WHO recommendations for the prevention and treatment of postpartum haemorrhage. World Health Organization. http://apps.who.int/iris/bitstream/10665/75411/1/9789241548502_eng.pdf. Accessed 7/29/13. • Smith JR, Brennan BG. Postpartum hemorrhage treatment & management. Medscape. http://emedicine.medscape.com/article/275038-treatment. Accessed 7/29/13. • Callaghan WM, et al. *Am J Obstet Gynecol.* 2010;202:353 e1.

Transfusions in postpartum women: Smith JR, Brennan BG. Postpartum hemorrhage. Medscape. http://emedicine.medscape.com/article/275038-overview. Accessed 7/29/13. • Callaghan WM, et al. *Am J Obstet Gynecol.* 2010;202:353 e1.

• Balki M, et al. *J Obstet Gynaecol Can.* 2008;30:1002. • Reyal F, et al. *Eur J Obstet Gynecol Reprod Biol.* 2004;112:61.

Death from postpartum bleeding: Kassebaum NJ, et al. *Lancet.* 2014;384:980. • Pregnancy mortality surveillance system. Centers for Disease Control and Prevention. http://www.cdc.gov/reproductivehealth/maternalinfanthealth/pmss .html. Accessed 8/30/13.

Deaths from clots versus bleeding: Deep vein thrombosis (DVT)/Pulmonary embolism (PE)—Blood clot forming in a vein. Data & statistics. Centers for Disease Control and Prevention. http://www.cdc.gov/ncbddd/dvt/data.html. Accessed 8/30/13.

Treating high cholesterol: Eaton SB. *Lipids.* 1992;27:814. • Anderson JL. In: Goldman L, Schafer AI, eds. *Goldman-Cecil Medicine;* 2016 (see bibliography). • O'Keefe JH Jr., Cordain L. *Mayo Clin Proc.* 2004;79:101. • O'Keefe JH Jr., et al. *J Am Coll Cardiol.* 2004;43:2142.

Blood thinners after a heart attack: Schulman S, Hirsh J. In: Goldman L, Schafer AI, eds. *Goldman-Cecil Medicine;* 2016 (see bibliography).

Percent of strokes from clotting: Hata J, Kiyohara Y. *Circ J.* 2013;77:1923. • Yates PO. *Lancet.* 1964;1:65. • Hilmarsson A, et al. *Stroke.* 2013;44:1714. • Arboix A, et al. *Cerebrovasc Dis.* 2008;26:509. • Goldstein LB. In: Goldman L, Schafer AI, eds. *Goldman-Cecil Medicine;* 2016 (see bibliography).

Saving brain tissue after a stroke: Goldstein LB. In: Goldman L, Schafer AI, eds. *Goldman-Cecil Medicine;* 2016 (see bibliography). • Mazighi M, et al. *Circulation.* 2013;127:1980. • Saver JL, et al. *JAMA.* 2013;309:2480.

Carotid endarterectomy: Goldstein LB. In: Goldman L, Schafer AI, eds. *Goldman-Cecil Medicine;* 2016 (see bibliography).

Potential new technologies for stroke: Berkhemer OA, et al. *N Engl J Med.* 2015;372:11. • Campbell BC, et al. *N Engl J Med.* 2015;372:1009. • Goyal M, et al. *N Engl J Med.* 2015;372:1019.

Chris Staniforth: Carollo K. Man dies from blood clot after marathon gaming. ABC News. http://abcnews.go.com/Health/extreme-gamer-dies-pulmonary -embolism/story?id=14212015. Accessed 8/5/11.

Preventing blood clots on planes: Watson HG, Baglin TP. *Br J Haematol.* 2011;152:31. • Philbrick JT, et al. *J Gen Intern Med.* 2007;22:107. • Jacobson BF, et al. *S Afr Med J.* 2003;93:522.

Deaths from clotting versus bleeding: Murphy SL, et al. Deaths: Final data for 2010. National Vital Statistics Reports. National Center for Health Statistics. http://www.cdc.gov/nchs/data/nvsr/nvsr61/nvsr61_04.pdf. Accessed 8/30/13.

• Deep vein thrombosis (DVT)/Pulmonary embolism (PE)—Blood clot forming in a vein. Data & statistics. Centers for Disease Control and Prevention. http://www.cdc.gov/ncbddd/dvt/data.html. Accessed 8/30/13.

Clotting versus bleeding as causes of maternal death: Berg CJ, et al. *Obstet Gynecol.* 2011;117:1230.

High blood pressure versus postpartum hemorrhage: Kassebaum NJ, et al. *Lancet.* 2014;384:980.

Chapter 6: Can Our Genes Evolve Fast Enough to Solve Our Problems?

Life expectancy: Olshansky SJ, et al. *Health Aff (Millwood).* 2012;31:1803. • Naghavi M, et al. *Lancet.* 2015;385:117.

Perpetuating the species: Dobson AP, Lyles AM. *Conserv Biol.* 1989;3:362.

Gambian, Ache, and Dogon live births: Allal N, et al. *Proc Biol Sci.* 2004;271:465. • Strassmann BI, Gillespie B. *Proc Biol Sci.* 2002;269:553.

Human and Neanderthal life expectancy: Trinkaus E. *J Archaeol Sci.* 1995;22:121. • Gurven M, Kaplan H. *Popul Dev Rev.* 2007;33:321. • Finch CE. *Proc Natl Acad Sci U S A.* 2010;107 Suppl 1:1718.

Benefit of grandmothers: Gurven M, Kaplan H. *Popul Dev Rev.* 2007;33:321.

Paleolithic hunting skills: Walker R, et al. *J Hum Evol.* 2002;42:639. • Gurven M, et al. *J Hum Evol.* 2006;51:454.

Interval between children: Knodel J. *Popul Stud (Camb).* 1968;22:297.

Historical life expectancy: Gurven M, Kaplan H. *Popul Dev Rev.* 2007;33:321. • Finch CE. *Proc Natl Acad Sci U S A.* 2010;107 Suppl 1:1718. • Life expectancy by age, 1850–2011. Infoplease. http://www.infoplease.com/ipa/A0005140.html. Accessed 6/12/15.

Improvement in life expectancy: Easterlin RA. *J Econ Perspect.* 2000;14:7. • Mulbrandon C. Last 2,000 years of growth in world income and population (revised). Data source: Angus Maddison, University of Groningen. Visualizing Economics. http://visualizingeconomics.com/blog/2007/11/21/last-2000-of-growth-in-world-income-and-population-revised. Accessed 4/19/12.

Current life expectancy and childhood deaths: Naghavi M, et al. *Lancet.* 2015;385:117. • Wang H, et al. *Lancet.* 2014;384:957.

Longest life expectancies: Naghavi M, et al. *Lancet.* 2015;385:117. • Fuchs VR. *JAMA.* 2013;309:33.

European height: Easterlin RA. *J Econ Perspect.* 2000;14:7. • Fogel RW. *Hist Methods.* 1993;26:5.

Worldwide mortality rates: Murray CJ, et al. *Lancet.* 2012;380:2197. • Lozano R, et al. *Lancet.* 2012;380:2095. • Murray CJ, et al. *Lancet.* 2012;380:2063.

• Ezzati M, Riboli E. *N Engl J Med.* 2013;369:954. • Murray CJ, Lopez AD. *N Engl J Med.* 2013;369:448.

Suicide deaths and attempts: Deaths: Final data for 2012. National Vital Statistics Reports. National Center for Health Statistics. http://www.cdc.gov/nchs/data/nvsr/nvsr63/nvsr63_09.pdf. Accessed 2/24/15. • The DAWN (Drug Abuse Warning Network) Report. Emergency department visits for drug-related suicide attempts among middle-aged adults aged 45 to 64. Substance Abuse and Mental Health Services Administration. http://www.samhsa.gov/data/2K14/DAWN154/sr154-suicide-attempts-2014.htm. Accessed 8/13/14.

US life expectancy by education: Wang H, et al. *Popul Health Metr.* 2013;11:8.

Future diabetes in India and China: Lipska K. The global diabetes epidemic. *New York Times.* http://nyti.ms/1k07qeE. Accessed 1/25/15.

Mortality projections: Mathers CD, Loncar D. *PLoS Med.* 2006;3:e442. • Calle EE, et al. *N Engl J Med.* 2003;348:1625.

Natural selection and reproductive success: Ackermann M, Pletcher SD. In: Stearns SC, Koella JC, eds. *Evolution in Health and Disease;* 2008 (see bibliography).

Evolution of depression: Keedwell P. *How Sadness Survived;* 2008 (see bibliography).

Reproductive success in obese women: Jacobsen BK, et al. *J Womens Health (Larchmt).* 2013;22:460. • Frisco ML, Weden M. *J Marriage Fam.* 2013;75:920. • Jacobsen BK, et al. *Eur J Epidemiol.* 2012;27:923. • Bellver J, et al. *Fertil Steril.* 2013;100:1050. • Boots C, Stephenson MD. *Semin Reprod Med.* 2011;29:507. • Aune D, et al. *JAMA.* 2014;311:1536. • Cnattingius S, et al. *JAMA.* 2013;309:2362. • Reynolds RM, et al. *BMJ.* 2013;347:f4539.

Obese women perhaps no longer lagging: Frisco ML, et al. *Soc Sci Med.* 2012;74:1703.

Obesity in offspring of obese parents: Jacobson P, et al. *Am J Epidemiol.* 2007;165:101.

Gene flow and genetic drift: Andrews CA. *Nature Education Knowledge.* 2010;3:5.

Y chromosomes in Mexico City: Martinez-Marignac VL, et al. *Hum Genet.* 2007;120:807.

Worldwide human mutations: Keinan A, Clark AG. *Science.* 2012;336:740. • Hawks J, et al. *Proc Natl Acad Sci U S A.* 2007;104:20753.

Siberian fat metabolism: *Science.* 2013;339:496.

Chance a single advantageous mutation will spread: Otto SP, Whitlock MC. *Genetics.* 1997;146:723.

Accumulation of random neutral mutations: Molecular clocks. University of California Museum of Paleontology. http://evolution.berkeley.edu/evosite/

evo101/IIE1cMolecularclocks.shtml. Accessed 3/27/13. • Brain M. How carbon-14 dating works. HowStuffWorks. http://science.howstuffworks.com/environmental/ earth/geology/carbon-14.htm. Accessed 3/27/13.

Mitochondrial Eve: Molecular clocks. University of California Museum of Paleontology. http://evolution.berkeley.edu/evosite/evo101/IIE1cMolecular clocks.shtml. Accessed 3/27/13. • Cann RL, et al. *Nature.* 1987;325:31. • Cyran KA, Kimmel M. *Theor Popul Biol.* 2010;78:165. • Poznik GD, et al. *Science.* 2013;341:562.

Common male Y chromosome: Poznik GD, et al. *Science.* 2013;341:562. • Cruciani F, et al. *Am J Hum Genet.* 2011;88:814. • Mendez FL, et al. *Am J Hum Genet.* 2013;92:454. • Gibbons A. *Science.* 2012;338:189. • Francalacci P, et al. *Science.* 2013;341:565.

Polymorphisms and their accumulation: Making SNPs make sense. Genetic Science Learning Center, University of Utah. http://learn.genetics.utah.edu/ content/pharma/snips/. Accessed 9/9/13. • Wang J, et al. *Nature.* 2008;456:60. • Wheeler DA, et al. *Nature.* 2008;452:872. • Ng PC, et al. *PLoS Genet.* 2008;4:e1000160. • Takahata N. *Genetics.* 2007;176:1.

Malaria and balanced selection: Steinberg MH. In: Goldman L, Schafer AI, eds. *Goldman-Cecil Medicine;* 2016 (see bibliography).

Chronic wasting disease: Belay ED, et al. *Emerg Infect Dis.* 2004;10:977.

Kuru: Lindenbaum S. *Philos Trans R Soc Lond B Biol Sci.* 2008;363:3715. • Whitfield JT, et al. *Philos Trans R Soc Lond B Biol Sci.* 2008;363:3721.

Prion configuration: Bosque PJ. In: Goldman L, Schafer AI, eds. *Goldman-Cecil Medicine;* 2016 (see bibliography).

Prions, kuru, and mad cow disease: Bishop MT, et al. *BMC Med Genet.* 2009;10:146. • Mead S, et al. *Science.* 2003;300:640. • Mead S, et al. *Philos Trans R Soc Lond B Biol Sci.* 2008;363:3741. • Diack AB, et al. *Prion.* 2014;8:286. • Saba R, Booth SA. *Public Health Genomics.* 2013;16:17. • Mackay GA, et al. *Int J Mol Epidemiol Genet.* 2011;2:217.

New G127V mutation: Mead S, et al. *N Engl J Med.* 2009;361:2056.

Why no mad sheep disease: Concepcion GP, et al. *Med Hypotheses.* 2005;64:919.

Projected human population and time to spread advantageous mutations: Rosenberg M. Current world population and world population growth since the year one. http://geography.about.com/od/obtainpopulationdata/a/ worldpopulation.htm. Accessed 4/19/12. • Aubuchon V. World population growth history. Vaughn's summaries. http://www.vaughns-1-pagers.com/history/ world-population-growth.htm. Accessed 4/19/12.

Black Death (plague): World population: historical estimates of world population. http://www.census.gov/population/international/data/worldpop/table_history.php. United States Census Bureau. Accessed 8/5/13. • Biraben JN, Langaney A. *An essay concerning mankind's evolution. Population,* selected papers 4. Paris: National Institute for Population Studies; 1980.

European diseases introduced to North America: The story of . . . smallpox — and other deadly Eurasian germs. PBS. http://www.pbs.org/gunsgermssteel/variables/smallpox.html. Accessed 8/13/12.

Spread of a polymorphism already in 10 percent of the population: Relethford J. *Human Population Genetics;* 2012 (see bibliography).

Lethal recessive mutation carried in 30 percent of population: Gao Z, et al. *Genetics.* 2015;199:1243.

Neutral mutations become disadvantageous: Casals F, Bertranpetit J. *Science.* 2012;337:39.

Origin of Tay-Sachs mutation: Karpati M, et al. *Neurogenetics.* 2004;5:35.

Equilibrium and fatal homozygous mutations: Relethford J. *Human Population Genetics;* 2012 (see bibliography).

Genetics of nearsightedness: Tran-Viet KN, et al. *Am J Hum Genet.* 2013;92:820. • Jones LA, et al. *Invest Ophthalmol Vis Sci.* 2007;48:3524.

Environmental causes of nearsightedness: Guggenheim JA, et al. *Invest Ophthalmol Vis Sci.* 2012;53:2856. • Wu PC, et al. *Ophthalmic Epidemiol.* 2010;17:338. • Dirani M, et al. *Br J Ophthalmol.* 2009;93:997. • Rose KA, et al. *Arch Ophthalmol.* 2008;126:527. • He M, et al. *JAMA.* 2015;314:1142.

Jean-Baptiste Lamarck: Lamarck JB. *Zoological Philosophy : An Exposition with Regard to the Natural History of Animals.* New York: Hafner Publishing Company; 1963.

Trofim Lysenko: Soyfer V. *Lysenko and the Tragedy of Soviet Science.* New Brunswick, NJ: Rutgers University Press; 1994.

Epigenetic DNA tags: Epigenetics and inheritance. Genetic Science Learning Center, University of Utah. http://learn.genetics.utah.edu/content/epigenetics/inheritance/. Accessed 8/14/12. • Barrès R, et al. *Cell Metab.* 2012;15:405. • Rönn T, et al. *PLoS Genet.* 2013;9:e1003572.

Human epigenome and its inheritance: The epigenome learns from its experiences. Genetic Science Learning Center, University of Utah. http://learn.genetics.utah.edu/content/epigenetics/epi_learns/. Accessed 8/14/12. • Geoghegan JL, Spencer HG. *Theor Popul Biol.* 2013;88C:9. • White YA, et al. *Nat Med.* 2012;18:413. • Grossniklaus U, et al. *Nat Rev Genet.* 2013;14:228. • Lim JP, Brunet A. *Trends Genet.* 2013;29:176. • Duffié R, Bourc'his D. *Curr Top Dev Biol.* 2013;104:293. • Stringer

JM, et al. *Reproduction*. 2013;146:R37. • Hackett JA, et al. *Science*. 2013;339:448. • Guibert S, et al. *Genome Res*. 2012;22:633. • Lane M, et al. *Science*. 2014;345:756.

Father's obesity influences children's sugar metabolism: Soubry A, et al. *BMC Med*. 2013;11:29. • Ost A, et al. *Cell*. 2014;159:1352. • Ozanne SE. *N Engl J Med*. 2015;372:973.

Nonhuman epigenetic inheritance: Pennisi E. *Science*. 2013;341:1055. • Jablonka E. *Clin Pharmacol Ther*. 2012;92:683.

Grandparents' nutrition influencing children's obesity and diabetes: Rando OJ. *Cell*. 2012;151:702. • Maron DF. Diet during pregnancy linked to diabetes in grandchildren. *Scientific American*. http://www.scientificamerican.com/article/diet-during-pregnancy-linked-to-diabetes-in-grandchildren/. Accessed 8/19/14.

Quebec study of rural and urban: Willführ KP, Myrskylä M. *Am J Hum Biol*. 2013;25:318.

Influence of mother's malnutrition or smoking on child's obesity: Ravelli GP, et al. *N Engl J Med*. 1976;295:349. • Moller SE, et al. *PLoS One*. 2014;9:e109184.

Epigenetics of environmental toxins and cancer: Guerrero-Bosagna C, et al. *PLoS One*. 2013;8:e59922. • Skinner MK, et al. *PLoS One*. 2013;8:e66318. • *N Engl J Med*. 2013;368:2059.

Chapter 7: Changing Our Behavior

Oprah Winfrey weight gain and loss: Haupt A. Celebrity weight loss: tales of the scales. *US News & World Report*. http://t.usnews.com/s208. Accessed 3/12/14. • Rousseau C. Oprah Winfrey 'embarrassed' at weight gain. *The Independent*. http://ind.pn/1LDpqKD. Accessed 3/13/14. • Rosen M. Oprah overcomes. 1994. *People*. http://www.people.com/people/article/0,,20107260,00.html. Accessed 3/13/14. • Winfrey O. "How Did I Let This Happen Again?" Oprah.com. http://www.oprah.com/spirit/Oprahs-Battle-with-Weight-Gain-O-January-2009-Cover. Accessed 3/13/14. • Silverman S. Oprah Winfrey admits to tipping the scales at 200 lbs. *People*. http://www.people.com/people/article/0,,20245089,00.html. Accessed 3/13/14. • Duggan D. Oprah's battle with weight loss. 2010. Body + Soul. http://www.bodyandsoul.com.au/weight+loss/lose+weight/oprahs+battle+with+weight+loss,9787. Accessed 3/13/14.

New Year's resolutions: Williams R. Why New Year's resolutions fail. *Psychology Today*. http://www.psychologytoday.com/blog/wired-success/201012/why-new-years-resolutions-fail. Accessed 3/13/14.

Planning better: Allan JL, et al. *Ann Behav Med*. 2013;46:114.

Implementation intentions: Gollwitzer PM, Sheeran P. *Adv Exp Soc Psychol*. 2006;38:69.

Executive control: Allan JL, et al. *Psychol Health.* 2011;26:635.

Achieving goals: Milkman KL, Volpp KG. How to keep your resolutions. *New York Times.* http://nyti.ms/1khpl56. Accessed 1/6/14.

The downside of perfectionism: Powers TA, et al. *Pers Soc Psychol Bull.* 2005;31:902.

Strong habits: Webb TL, et al. *Br J Soc Psychol.* 2009;48:507.

Situational cues: Van't Riet J, et al. *Appetite.* 2011;57:585.

Unrealistic expectations: Kahneman D, Tversky A. Intuitive prediction: biases and corrective procedures. In: Makridakis S, Wheelwright SC, eds. *Forecasting : TIMS Studies in the Management Sciences.* Amsterdam, New York, and Oxford: North-Holland Publishing Company; 1979;12:313.

Succumbing to temptation: Kroese FM, et al. *Eat Behav.* 2013;14:522. • Taylor C, et al. *Br J Soc Psychol.* 2014;53:501.

Chocolate experiment: Taylor C, et al. *Br J Soc Psychol.* 2014;53:501.

Past predicts future: Gollwitzer PM, Sheeran P. *Adv Exp Soc Psychol.* 2006;38:69.

Weight loss in a sedentary person: Hall KD, et al. *Lancet.* 2011;378:826.

Calories determine your weight: Wadden TA, et al. *Circulation.* 2012;125:1157. • Walker TB, Parker MJ. *J Am Coll Nutr.* 2014;33:347. • Hollis JH, Mattes RD. *Curr Diab Rep.* 2005;5:374.

Low-carb versus Paleolithic diet: Bray GA, et al. *JAMA.* 2012;307:47. • Hall KD, et al. *Cell Metab.* 2015;22:427. • Bazzano LA, et al. *Ann Intern Med.* 2014;161:309. • Sacks FM, et al. *N Engl J Med.* 2009;360:859. • Wycherley TP, et al. *Am J Clin Nutr.* 2012;96:1281. • Larsen TM, et al. *N Engl J Med.* 2010;363:2102. • Bueno NB, et al. *Br J Nutr.* 2013;110:1178. • Riera-Crichton D, Tefft N. *Econ Hum Biol.* 2014;14:33. • Johnston BC, et al. *JAMA.* 2014;312:923. • LeCheminant JD, et al. *Nutr J.* 2007;6:36. • LeCheminant JD, et al. *Lipids Health Dis.* 2010;9:54. • Jensen MD, Ryan DH. *JAMA.* 2014;311:23.

Snacking mice: Hatori M, et al. *Cell Metab.* 2012;15:848.

Mediterranean diet: Esposito K, et al. *Diabetes Care.* 2014;37:1824. • Carter P, et al. *J Hum Nutr Diet.* 2014;27:280. • Ajala O, et al. *Am J Clin Nutr.* 2013;97:505. • Estruch R, et al. *N Engl J Med.* 2013;368:1279. • Chiva-Blanch G, et al. *Curr Atheroscler Rep.* 2014;16:446. • Dalen JE, Devries S. *Am J Med.* 2014;127:364. • Crous-Bou M, et al. *BMJ.* 2014;349:g6674.

Low-carb versus low-fat diets for metabolic health, diabetes, and cardiovascular disease: Hu T, et al. *Am J Epidemiol.* 2012;176 Suppl 7:S44. • Mirza NM, et al. *Int J Pediatr Obes.* 2011;6:e523. • Mirza NM, et al. *Am J Clin Nutr.* 2013;97:276. • Schwingshackl L, Hoffmann G. *Nutr J.* 2013;12:48. • de Souza

RJ, et al. *Am J Clin Nutr.* 2012;95:614. • Pereira MA, et al. *JAMA.* 2004;292:2482. • Makris AP, et al. *Obesity (Silver Spring).* 2011;19:2365. • Mann J, et al. *Lancet.* 2014;384:1479. • Lindstrom J, et al. *Diabetologia.* 2013;56:284.

Are some carbs better for you than others? Sacks FM, et al. *JAMA.* 2014;312:2531. • O'Connor A. Questioning the idea of good carbs, bad carbs. *New York Times.* http://nyti.ms/1uPw9Gq. Accessed 12/19/14.

Physician-supervised weight loss: Wadden TA, et al. *N Engl J Med.* 2011;365:1969. • Lin JS, et al. *Ann Intern Med.* 2014;161:568. • Wadden TA, et al. *JAMA.* 2014;312:1779.

Success of Nutrisystem, Jenny Craig, and Weight Watchers: Wadden TA, et al. *Circulation.* 2012;125:1157. • Gudzune KA, et al. *Ann Intern Med.* 2015;162:501.

Taft's weight loss: Levine DI. *Ann Intern Med.* 2013;159:565.

Success of weight-loss diets: Rossner SM, et al. *Obes Facts.* 2011;4:3. • Weiss EC, et al. *Am J Prev Med.* 2007;33:34.

Downside of yo-yo dieting: Brownell KD, et al. *Physiol Behav.* 1986;38:459.

Long-term approach to obesity: Casazza K, et al. *N Engl J Med.* 2013;368:446.

Implausibility of self-reported food intake: Archer E, et al. *PLoS One.* 2013;8:e76632.

Eating popcorn: Wansink B, Kim J. *J Nutr Educ Behav.* 2005;37:242.

Container, portion, and package size: Wansink B. *J Mark.* 1996;60:1. • Wansink B, Cheney MM. *JAMA.* 2005;293:1727. • Wansink B, et al. *Am J Prev Med.* 2006;31:240. • Fisher JO, Kral TV. *Physiol Behav.* 2008;94:39.

Secretaries eating candy: Wansink B, et al. *Int J Obes (Lond).* 2006;30:871.

Size of soup bowls: Wansink B, et al. *Obes Res.* 2005;13:93.

Food consumption in restaurants: Wansink B. *Annu Rev Nutr.* 2004;24:455.

Mindless eating: Moss M. The extraordinary science of addictive junk food. *New York Times Magazine.* http://nyti.ms/1FUm99j. Accessed 2/20/13. • Wansink B. *Mindless Eating : Why We Eat More Than We Think.* New York: Bantam Books; 2006.

Calorie-dense foods: Moss M. *Salt, Sugar, Fat : How the Food Giants Hooked Us.* New York: Random House; 2013.

Triple threat and what to do about it: Chandon P, Wansink B. *Nutr Rev.* 2012;70:571. • Wansink B. *Physiol Behav.* 2010;100:454.

Women's perception of ideal body size: Christakis NA, Fowler JH. *N Engl J Med.* 2007;357:370. • Martijn C, et al. *Health Psychol.* 2013;32:433.

Blair River and Heart Attack Grill: Johnson WB. 575-pound Heart Attack Grill spokesman dies at 29. *USA Today.* http://usatoday30.usatoday.com/news/nation/2011-03-04-restaurant-spokesman-dies_N.htm. Accessed 12/31/13.

Vouchers to move to different neighborhoods: Ludwig J, et al. *N Engl J Med.* 2011;365:1509.

Breanna Bond: Breanna Bond, 9, loses 66 pounds. *ABC News.* http://abc news.go.com/blogs/health/2012/12/10/breanna-bond-9-loses-66-pounds/. Accessed 4/24/14.

Jennifer Hudson: How much does Jennifer Hudson weigh? HelloBeautiful .com. http://hellobeautiful.com/2012/07/29/jennifer-hudson-weight/. Accessed 8/20/15. • Jennifer Hudson: "I'm prouder of my weight loss than my Oscar!" *Huffington Post.* http://www.huffingtonpost.com/2011/08/17/jenniferhudson-weight-loss_n_929188.html. Accessed 8/20/15.

Try harder: Thaler RH, Sunstein CR. *Nudge : Improving Decisions About Health, Wealth, and Happiness.* New Haven, CT: Yale University Press; 2008.

Decline in US cholesterol levels: Taubes G. Why the campaign to stop America's obesity crisis keeps failing. *Newsweek.* http://www.newsweek.com/why -campaign-stop-americas-obesity-crisis-keeps-failing-64977. Accessed 4/24/14.

Snacking mice and fruit flies: Hatori M, et al. *Cell Metab.* 2012;15:848. • Chaix A, et al. *Cell Metab.* 2014;20:991. • Reynolds G. A 12-hour window for a healthy weight. *New York Times.* http://nyti.ms/1ylMetT. Accessed 1/20/15. • Gill S, et al. *Science.* 2015;347:1265.

Detecting dietary sugars in blood: Callier V. *Science.* 2015;348:488. • Hedrick VE, et al. *Public Health Nutr.* 2015:1.

Obesity in schoolchildren: Tavernise S. Poor children show a decline in obesity rate. *New York Times.* http://nyti.ms/17vAJhK. Accessed 8/6/13. • Tavernise S. Obesity in young is seen as falling in several cities. *New York Times.* http://nyti.ms/ VvIZIz. Accessed 12/11/12. • Tavernise S. Children in U.S. are eating fewer calories, study finds. *New York Times.* http://nyti.ms/Yo1K1l. Accessed 2/21/13. • Ogden CL, et al. *JAMA.* 2014;311:806. • Skinner AC, Skelton JA. *JAMA Pediatr.* 2014;168:561.

Worldwide obesity: Ng M, et al. *Lancet.* 2014;384:766.

US diabetes and obesity: Geiss LS, et al. *JAMA.* 2014;312:1218. • Report: US obesity rates increased in just six states in 2013. *AMA Morning Rounds.* American Medical Association; September 5, 2014.

Exercise-induced weight loss: Wadden TA, et al. *Circulation.* 2012;125:1157.

Exercise to improve outcomes: Rottensteiner M, et al. *Med Sci Sports Exerc.* 2015;47:509.

Exercise to prevent clots: Naci H, Ioannidis JP. *BMJ.* 2013;347:f5577.

Exercise alters fat cells and muscle cells: Rönn T, et al. *PLoS Genet.* 2013;9:e1003572. • Barres R, et al. *Cell Metab.* 2012;15:405.

Physical activity, including at work: Roux L, et al. *Am J Prev Med.* 2008;35:578. • Church TS, et al. *PLoS One.* 2011;6:e19657. • Stamatakis E, et al. *Prev Med.* 2007;45:416.

US exercise versus guidelines: Buchner DM. In: Goldman L, Schafer AI, eds. *Goldman-Cecil Medicine;* 2016 (see bibliography).

Effect of lack of physical activity: Lee I-M, et al. *Lancet.* 2012;380:219.

Overexercising rodents: Roberts MD, et al. *Am J Physiol Regul Integr Comp Physiol.* 2013;304:R1024. • Roberts MD, et al. *J Physiol.* 2014;592:2119. • Kelly SA, et al. *Physiol Genomics.* 2014;46:593. • Kelly SA, et al. *Genetics.* 2012;191:643. • Meek TH, et al. *J Exp Biol.* 2009;212:2908.

Dopamine: Roberts MD, et al. *J Physiol.* 2014;592:2119. • Mathes WF, et al. *Behav Brain Res.* 2010;210:155. • Blum K, et al. *J Psychoactive Drugs.* 2012;44:134. • Wise RA. *Neurotox Res.* 2008;14:169. • Roberts MD, et al. *Physiol Behav.* 2012;105:661.

Exercise reduces depression: Cooney G, et al. *JAMA.* 2014;311:2432. • Agudelo LZ, et al. *Cell.* 2014;159:33.

Hyperactive mice and human ADHD: Waters RP, et al. *Brain Res.* 2013;1508:9.

Tailoring exercise: Stadler G, et al. *Am J Prev Med.* 2009;36:29.

Exercise interventions: Roux L, et al. *Am J Prev Med.* 2008;35:578. • Metcalf B, et al. *BMJ.* 2012;345:e5888. • Michie S, et al. *Health Psychol.* 2009;28:690. • Orrow G, et al. *BMJ.* 2012;344:e1389.

Downsides of more than 2.3 grams of sodium per day: Kalogeropoulos AP, et al. *JAMA Intern Med.* 2015;175:410.

Safety of three to six grams of sodium per day: O'Donnell M, et al. *N Engl J Med.* 2014;371:612.

Data suggesting no more than 2.3 grams of sodium per day: Strazzullo P, et al. *BMJ.* 2009;339:b4567. • Whelton PK, et al. *Circulation.* 2012;126:2880.

Percent above recommended salt intake: Sifferlin A. 90% of Americans eat too much salt. *Time.* http://time.com/3944545/sodium-heart/. Accessed 7/15/15.

Benefit of a 1.2-gram-per-day sodium reduction: Bibbins-Domingo K, et al. *N Engl J Med.* 2010;362:590.

Salt reduction programs in the UK and Finland: He FJ, MacGregor GA. *J Hum Hypertens.* 2009;23:363. • Laatiken T, et al. *Eur J Clin Nutr.* 2006;60:965.

Genetics of happiness: Stubbe JH, et al. *Psychol Med.* 2005;35:1581. • Nes RB, et al. *Twin Res Hum Genet.* 2010;13:312.

Control over happiness: Lyubomirsky S. *The How of Happiness : A Scientific Approach to Getting the Life You Want.* New York: Penguin Press; 2008. • Reis HT, et al. *Pers Soc Psychol Bull.* 2000;26:419.

Factors influencing happiness: Kahneman D, et al. *Well-Being : The Foundations of Hedonic Psychology.* New York: Russell Sage Foundation; 1999.

Happiness by sex, age, and income: Bartels M, Boomsma DI. *Behav Genet.* 2009;39:605. • Cooper C, et al. *Int J Geriatr Psychiatry.* 2011;26:608. • Boyce CJ, et al. *Psychol Sci.* 2010;21:471.

Happiness and longevity: Sadler ME, et al. *Twin Res Hum Genet.* 2011;14:249.

Cognitive behavioral therapy: Lynch D, et al. *Psychol Med.* 2010;40:9. • Deacon BJ, Abramowitz JS. *J Clin Psychol.* 2004;60:429. • McGrath CL, et al. *JAMA Psychiatry.* 2013;70:821.

Prior physician visits of suicide victims: Tingley K. The suicide detective. *New York Times Magazine.* http://nyti.ms/18fHdXl. Accessed 7/1/13.

Suicide in active-duty troops: Zoroya G. Suicides in the army decline sharply. *USA Today.* http://usat.ly/LifXOq. Accessed 2/3/15.

Accuracy of psychiatric diagnosis: Mitchell AJ, et al. *Lancet.* 2009;374:609. • Fernandez A, et al. *Gen Hosp Psychiatry.* 2012;34:227.

Improving attendance in psychotherapy: Oldham M, et al. *J Consult Clin Psychol.* 2012;80:928.

Cognitive behavioral therapy and quality of life: Tyrer P, et al. *Lancet.* 2014;383:219.

Who is to blame for childhood obesity: Barry CL, et al. *N Engl J Med.* 2012;367:389.

Effects of providing information on caloric content of food: Block JP, Roberto CA. *JAMA.* 2014;312:887. • Dumanovsky T, et al. *Am J Public Health.* 2010;100:2520. • Krieger JW, et al. *Am J Prev Med.* 2013;44:595. • Tandon PS, et al. *Am J Prev Med.* 2011;41:434. • Finkelstein EA, et al. *Am J Prev Med.* 2011;40:122. • Yamamoto JA, et al. *J Adolesc Health.* 2005;37:397. • Gerend MA. *J Adolesc Health.* 2009;44:84. • Namba A, et al. *Prev Chronic Dis.* 2013;10:E101.

Experience in Starbucks: Tyler A. Changing the food environment. *NYU Physician.* 2011:16.

Burger King's french fries: Strom S. Burger King introducing a lower-fat french fry. *New York Times.* http://nyti.ms/18mbwZ9. Accessed 9/24/13. • Jargon J. Burger King drops lower-calorie fry 'Satisfries.' *Wall Street Journal.* http://on.wsj.com/1ziu6AO. Accessed 6/3/2015. • Patton L. Burger King stores discontinue Satisfries as sales fizzle. *Bloomberg Business.* http://www.bloomberg.com/news/articles/2014-08-13/burger-king-stores-discontinue-satisfries-as-sales-fizzle. Accessed 6/3/2015.

Voluntary reduction in calories and sugary drinks: Strom S. Soda makers Coca-Cola, PepsiCo and Dr Pepper join in effort to cut Americans' drink calories. *New York Times.* http://nyti.ms/Y1UgZA. Accessed 9/29/14.

Using pricing to influence food purchases: Chandon P, Wansink B. *Nutr Rev.* 2012;70:571.

Cigarette and gasoline taxes: Brown HS, Karson S. *Health Econ.* 2013;22:741. • Levy DT, et al. *Am J Prev Med.* 2012;43:S179. • Coady MH, et al. *Am J Public Health.* 2013;103:e54. • Bento AM, et al. *Am Econ Rev.* 2009;99:667.

Potential of taxes to influence food choices: Mozaffarian D, et al. *JAMA.* 2014;312:889. • Wang YC, et al. *Health Affairs.* 2012;31:199. • *N Engl J Med.* 2012;367:1464.

Smoking reduction in the United States: Levy DT, et al. *Am J Prev Med.* 2012;43:S179. • Meyers DG, et al. *J Am Coll Cardiol.* 2009;54:1249.

Motorcycle helmets: Deutermann WV. Calculating lives saved by motorcycle helmets. National Center for Statistics and Analysis (NCSA) of the National Highway Traffic Safety Administration (NHTSA). http://www-nrd.nhtsa.dot .gov/Pubs/809861.pdf. Accessed 6/10/15.

Trans-fat bans: Angell SY, et al. *Ann Intern Med.* 2012;157:81. • Brownell KD, Pomeranz JL. *N Engl J Med.* 2014;370:1773. • Farley TA. *JAMA.* 2012;308:1093. • FDA takes step to remove artificial trans fats from processed foods. US Food and Drug Administration. http://www.fda.gov/NewsEvents/ Newsroom/PressAnnouncements/ucm451237.htm. Accessed 6/16/15.

Even the threat of bans may help: Unnevehr LJ, Jagmanaite E. *Food Policy.* 2008;33:497.

FDA and salt regulation: Lowering salt in your diet. US Food and Drug Administration. http://www.fda.gov/forconsumers/consumerupdates/ucm181577.htm. Accessed 2/25/12.

Salt levels in US processed and restaurant food: Jacobson MF, et al. *JAMA Intern Med.* 2013;173:1285.

Banning sugar-sweetened drinks: Zap C. "Anti-Bloomberg bill" passed in Mississippi. Yahoo news. http://news.yahoo.com/blogs/the-lookout/anti-bloomberg -bill-passed-mississippi-215548774--finance.html. Accessed 3/14/13. • Elbel B, et al. *N Engl J Med.* 2012;367:680. • Niederdeppe J, et al. *Am J Public Health.* 2013;103:e92.

China's salt reduction policies: Hvistendahl M. *Science.* 2014;345:1268.

Chapter 8: Changing Our Biology

Prescription medicines in the prior month: Gu Q, et al. *NCHS Data Brief.* 2010;42:1. • Health, United States, 2013: With special feature on prescription drugs. National Center for Health Statistics. http://www.cdc.gov/nchs/data/hus/hus13.pdf. Accessed 11/19/14.

American use of alternative medicine: Perlman A. In: Goldman L, Schafer AI, eds. *Goldman-Cecil Medicine;* 2016 (see bibliography).

Steve Jobs's cancer: Walton AG. Steve Jobs' cancer treatment regrets. *Forbes.* http://onforb.es/rcmSaD. Accessed 7/23/14.

AMA declares obesity a disease: Fitzgerald K. Obesity is now a disease, American Medical Association decides. *Medical News Today.* http://www.medical newstoday.com/articles/262226.php. Accessed 7/7/14. • Hoyt CL, Burnette JL. Should obesity be a "disease"? *New York Times.* http://nyti.ms/1eelWeu. Accessed 7/7/14. • Pollack A. A.M.A. recognizes obesity as a disease. *New York Times.* http://nyti.ms/11ml6KK. Accessed 7/7/14.

When doctors recommend diet medications: Colman E. *Circulation.* 2012;125:2156. • Kushner RF. *Circulation.* 2012;126:2870.

FDA-approved diet medications: Yanovski SZ, Yanovski JA. *JAMA.* 2014;311:74. • *JAMA.* 2014;312:955. • Phillip A. Meet the newest FDA-approved prescription weight-loss drug: Contrave. *Washington Post.* http://wapo.st/1nNOzoL. Accessed 11/14/14. • Pollack A. New drug to treat obesity gains approval by F.D.A. *New York Times.* http://nyti.ms/1pR1p5c. Accessed 9/22/14. • FDA approves weight-management drug Saxenda. US Food and Drug Administration. http://www .fda.gov/NewsEvents/Newsroom/PressAnnouncements/ucm427913.htm. Accessed 1/30/15.

Orlistat and Qsymia: Yanovski SZ, Yanovski JA. *JAMA.* 2014;311:74.

Contrave benefits and side effects: Caixas A, et al. *Drug Des Devel Ther.* 2014;8:1419. • Verpeut JL, Bello NT. *Expert Opin Drug Saf.* 2014;13:831.

Use of liraglutide: Saxenda Medication Guide. Novo Nordisk. http:// www.novo-pi.com/saxenda_med.pdf. Accessed 7/14/15 • Pi-Sunyer X, et al. *N Eng J Med.* 2015;373:11.

Contrave and dopamine: Tellez LA, et al. *Science.* 2013;341:800.

Adiponectin: Holland WL, Scherer PE. *Science.* 2013;342:1460.

Brain stimulation for hunger: Cai H, et al. *Nat Neurosci.* 2014;17:1240.

Light to influence taste buds: Yarmolinsky DA, et al. *Cell.* 2009;139:234.

Obesogens: Thayer KA, et al. *Environ Health Perspect.* 2012;120:779.

Effects of central heating: Reynolds G. Let's cool it in the bedroom. *New York Times.* http://nyti.ms/1zL2u7n. Accessed 11/19/14.

Number of fat cells: Hall KD, et al. *Am J Clin Nutr.* 2012;95:989. • Efrat M, et al. *J Pediatr Endocrinol Metab.* 2013;26:197.

Brown fat: Doheny K. The truth about fat. WebMD. http://www.webmd .com/diet/features/the-truth-about-fat. Accessed 8/21/14. • Rogers NH. *Ann Med.* 2014:1.

Beige fat: Grens K. Activating beige fat. *The Scientist.* http://www.the -scientist.com/?articles.view/articleNo/40147/title/Activating-Beige-Fat/. Accessed 7/7/14.

Brown fat and room temperature: Lee P, et al. *Diabetes.* 2014;63:3686. • Moyer MW. Supercharging brown fat to battle obesity. *Scientific American.* http:// www.scientificamerican.com/article/supercharging-brown-fat-to-battle -obesity/. Accessed 8/21/14.

Exercise to make white fat act like brown fat: Roberts LD, et al. *Cell Metab.* 2014;19:96. • Rao RR, et al. *Cell.* 2014;157:1279. • Irving BA, et al. *Curr Obes Rep.* 2014;3:235.

FTO gene induces browning of fat cells: Claussnitzer M, et al. *N Engl J Med.* 2015. doi:10.1056/NEJMoa1502214.

Beta-aminoisobutyric acid: Roberts LD, et al. *Cell Metab.* 2014;19:96.

"Browning": Fang S, et al. *Nat Med.* 2015;21:159. • Fikes BJ. Diet drug fools the gut. *San Diego Union-Tribune.* http://www.utsandiego.com/news/2015/jan/ 05/diet-drug-salk-imaginary-meal-ron-evans/. Accessed 1/6/15. • Qiu Y, et al. *Cell.* 2014;157:1292.

Human microbiome: Turnbaugh PJ, et al. *Nature.* 2006;444:1027. • Tilg H, Kaser A. *J Clin Invest.* 2011;121:2126.

Effects of penicillin on microbiome and weight: Cox LM, et al. *Cell.* 2014;158:705. • Jess T. *N Engl J Med.* 2014;371:2526. • Bailey LC, et al. *JAMA Pediatr.* 2014;168:1063.

Fecal transplantation in mice: Backhed F, et al. *Proc Natl Acad Sci U S A.* 2004;101:15718.

How microbiome affects weight: Furlow B. *Lancet Diabetes Endocrinol.* 2013;1 Suppl 1:s4. • Chen Z, et al. *J Clin Invest.* 2014;124:3391.

Artificial sweeteners and antibiotics: Suez J, et al. *Nature.* 2014;514:181. • Chang K. Artificial sweeteners may disrupt body's blood sugar controls. *New York Times.* http://nyti.ms/Xjn3bi. Accessed 9/22/14.

Christensenellaceae: Pennisi E. *Science.* 2014;346:687.

Treating obesity in mice: Pennisi E. *Science.* 2014;346:687. • Nieuwdorp M, et al. *Gastroenterology.* 2014;146:1525.

Fecal transplantation and **Clostridium difficile:** Van Nood E, et al. *N Engl J Med.* 2013;368:407.

Stomach reduction surgery: O'Brien PE, et al. *JAMA.* 2010;303:519. • Chakravarty PD, et al. *Surgeon.* 2012;10:172. • Courcoulas AP, et al. *JAMA Surg.* 2014;149:707. • Schauer PR, et al. *N Engl J Med.* 2014;370:2002. • Schauer PR, et al. *N Engl J Med.* 2012;366:1567. • Mingrone G, et al. *N Engl J Med.* 2012;366:1577. • Sjöström L, et al. *JAMA.* 2014;311:2297. • Ikramuddin S, et al. *JAMA.* 2013;309:2240. • Saeidi N, et al. *Science.* 2013;341:406. • Arterburn DE, Courcoulas AP. *BMJ.* 2014;349:g3961. • Liou AP, et al. *Sci Transl Med.* 2013;5:178ra41.

Weight-loss surgery in a two-year-old: Mohaidly MA, et al. *Int J Surg Case Rep.* 2013;4:1057.

Paul Mason: Lyall S. One-third the man he used to be, and proud of it. *New York Times.* http://nyti.ms/VI5vnQ. Accessed 7/7/14.

Governor Chris Christie: Zernike K, Santora M. Weight led governor to surgery. *New York Times.* http://nyti.ms/12aozes. Accessed July 7/7/14. • Mucha P. Gov. Christie spells out weight loss. Philly.com. http://www.philly.com/philly/news/politics/Gov_Christie_spells_out_weight_loss_.html. Accessed 9/22/14.

Weight loss via expandable balloons, expandable capsules, vagal stimulation, and aspiration therapy: Dixon JB, et al. *Circulation.* 2012;126:774. • Pollack A. Early results arrive on weight-loss pills that expand in the stomach. *New York Times.* http://nyti.ms/1lioVLd. Accessed 7/7/14. • Hand L. FDA panel mixed on implanted weight-loss device. Medscape. http://www.medscape.com/viewarticle/826946. Accessed 7/7/14. • Sullivan S, et al. *Gastroenterology.* 2013;145:1245. • Saint Louis C. F.D.A. approves surgical implant to treat obesity. *New York Times.* http://nyti.ms/1zf77Iy. Accessed 1/15/15.

SLC30A8 gene: Flannick J, et al. *Nat Genet.* 2014;46:357.

Lazy mice don't run more: Roberts MD, et al. *Physiol Behav.* 2012;105:661.

Mimicking exercise to treat depression: Reynolds G. How exercise may protect against depression. *New York Times.* http://nyti.ms/YK7r1K. Accessed 10/6/14. • Agudelo LZ, et al. *Cell.* 2014;159:33.

Blood pressure medicines: Victor RG. In: Goldman L, Schafer AI, eds. *Goldman-Cecil Medicine;* 2016 (see bibliography). • Jaffe MG, et al. *JAMA.* 2013;310:699.

Treating renal artery stenosis: Cooper CJ, et al. *N Engl J Med.* 2014;370:13.

Failure of radio-frequency kidney nerve ablation: Bhatt DL, et al. *N Engl J Med.* 2014;370:1393.

Future of directed blood pressure treatment: Laurent S, et al. *Lancet.* 2012;380:591.

How medications for depression and anxiety work: Krishnan V, Nestler EJ. *Am J Psychiatry.* 2010;167:1305.

SSRIs and SNRIs: Lyness JM. In: Goldman L, Schafer AI, eds. *Goldman-Cecil Medicine;* 2016 (see bibliography). • Arroll B, et al. *Cochrane Database Syst Rev.* 2009:CD007954.

Electroconvulsive therapy: Lyness JM. In: Goldman L, Schafer AI, eds. *Goldman-Cecil Medicine;* 2016 (see bibliography).

Benefit of continuing antidepressant medication: Borges S, et al. *J Clin Psychiatry.* 2014;75:205.

Anti-anxiety medication: Lyness JM. In: Goldman L, Schafer AI, eds. *Goldman-Cecil Medicine;* 2016 (see bibliography). • Anxiety disorders. Bethesda, MD: National Institute of Mental Health, National Institutes of Health, US Department of Health and Human Services. NIH publication 09 3879; 2009.

Medications for PTSD: Fazel S, et al. *Lancet.* 2014;384:1206.

Extinction behavior training: Karpova NN, et al. *Science.* 2011;334:1731.

Ketamine: Murrough JW, et al. *Am J Psychiatry.* 2013;170:1134. • Piroli GG, et al. *Exp Neurol.* 2013;241:184. • Feder A, et al. *JAMA Psychiatry.* 2014;71:681. • Lapidus KA, et al. *Biol Psychiatry.* 2014;76:970.

GABA in anxiety and depression: Mohler H. *Neuropharmacology.* 2012;62:42.

Deep brain stimulation for depression: Morishita T, et al. *Neurotherapeutics.* 2014;11:475.

Optogenetics: Deisseroth K. *Sci Am.* 2010;303:48.

Conditioning mice to have good memories: Redondo RL, et al. *Nature.* 2014;513:426. • Williams R. Light-activated memory switch. *The Scientist.* http://www.the-scientist.com/?articles.view/articleNo/40889/title/Light-Activated-Memory-Switch/. Accessed 11/13/14.

Fiber-optic cables: Belluck P. Using light technique, scientists find dimmer switch for memories in mice. *New York Times.* http://nyti.ms/1AUKrvl. Accessed 8/29/14.

Aspirin and salicylic acid: Awtry EH, Loscalzo J. *Circulation.* 2000;101:1206. • Stone E. *Phil Trans.* 1763;53:195. • Roth GJ, Majerus PW. *J Clin Invest.* 1975;56:624. • Rosenkranz B, et al. *Br J Clin Pharmacol.* 1986;21:309.

Aspirin to prevent heart attack and stroke: Schulman S, Hirsh J. In: Goldman L, Schafer AI, eds. *Goldman-Cecil Medicine;* 2016 (see bibliography).

Antiplatelet medications after coronary artery stents: Teirstein PS, Lytle BW. In: Goldman L, Schafer AI, eds. *Goldman-Cecil Medicine;* 2016 (see bibliography).

Atrial fibrillation: Garan H. In: Goldman L, Schafer AI, eds. *Goldman-Cecil Medicine;* 2016 (see bibliography).

Warfarin: Song Y, et al. *Curr Biol.* 2011;21:1296.

New drugs for atrial fibrillation: Schulman S, Hirsh J. In: Goldman L, Schafer AI, eds. *Goldman-Cecil Medicine;* 2016 (see bibliography).

Warfarin for venous clots: Ginsberg J. In: Goldman L, Schafer AI, eds. *Goldman-Cecil Medicine;* 2016 (see bibliography).

Number of Americans with coronary artery disease, stroke, or atrial fibrillation: What is atrial fibrillation (AFib or AF)? American Heart Association. http://www.heart.org/HEARTORG/Conditions/Arrhythmia/AboutArrhythmia/What-is-Atrial-Fibrillation-AFib-or-AF_UCM_423748_Article.jsp. Accessed 7/8/14. • Survivors. National Stroke Association. http://www.stroke.org/site/PageServer?pagename=surv. Accessed 7/8/14. • *Morb Mortal Wkly Rep.* 2011;60:1377.

Future anticlotting possibilities: Moeckel D, et al. *Sci Transl Med.* 2014;6:248ra105.

Who should take aspirin: Halvorsen S, et al. *J Am Coll Cardiol.* 2014;64:319.

Aspirin and preeclampsia: Henderson JT, et al. *Ann Intern Med.* 2014;160:695. • LeFevre ML. *Ann Intern Med.* 2014;161:819.

Who should take statins: Stone NJ, et al. *Circulation.* 2014;129:S1.

Ezetimibe: *N Engl J Med.* 2014;371:2072. • Kolata G. Study finds alternative to anti-cholesterol drug. *New York Times.* http://nyti.ms/1xeAZCg. Accessed 11/18/14. • Cannon CP, et al. *N Engl J Med.* 2015;372:2387.

Drugs to block PCSK9: Raal FJ, et al. *Lancet.* 2015;385:331. • Sabatine MS, et al. *N Engl J Med.* 2015;372:1500. • Robinson JG, et al. *N Engl J Med.* 2015;372:1489.

Benefits of APOC3 mutations: Jong MC, et al. *J Lipid Res.* 2001;42:1578. • Baroukh NN, et al. Lawrence Berkeley National Laboratory; 2003. Retrieved from: http://www.escholarship.org/uc/item/4cp3d5zc. Accessed 7/7/14. • Crosby J, et al. *N Engl J Med.* 2014;371:22. • Jorgensen AB, et al. *N Engl J Med.* 2014;371:32.

Mitt Romney's medications: Pittman D. Romney takes daily statin, aspirin. MedPage Today. http://www.medpagetoday.com/Washington-Watch/Election Coverage/34908. Accessed 7/7/14.

Altering or repairing DNA or RNA: Reenan R. *N Engl J Med.* 2014;370:172. • Rosenberg SM, Queitsch C. *Science.* 2014;343:1088. • Naryshkin NA, et al. *Science.* 2014;345:688. • Long C, et al. *Science.* 2014;345:1184. • Ding Q, et al. *Circ Res.* 2014;115:488. • Liang P, et al. *Protein Cell.* 2015;6:363. • Kolata G. Chinese scientists edit genes of human embryos, raising concerns. *New York Times.* http://nyti.ms/1PqeCS4. Accessed 4/24/15. • Baltimore D, et al. *Science.* 2015;348:36. • Wade N. Scientists seek ban on method of editing the human genome. *New York Times.* http://nyti.ms/19DfnrB. Accessed 3/20/15.

Epigenetic changes to alter gene function: Libri V, et al. *Lancet.* 2014;384:504. • Pennisi E. *Science.* 2015;348:618. • Hoffman Y, Pilpel Y. *Science.* 2015;348:41. • Popp MW, Maquat LE. *Science.* 2015;347:1316.

Making babies like the taste of garlic: Underwood E. *Science.* 2014;345:750.

Human trials to edit RNA: Haussecker D, Kay MA. *Science.* 2015;347:1069.

Mimicking unusual beneficial mutations: Kolata G. In a new approach to fighting disease, helpful genetic mutations are sought. *New York Times.* http://nyti .ms/1y02HEY. Accessed 1/5/15. • Kaiser J. *Science.* 2014;344:687.

New HIV therapies: Tebas P, et al. *N Engl J Med.* 2014;370:901. • Gardner MR, et al. *Nature.* 2015;519:87.

Editing DNA: Wei C, et al. *J Genet Genomics.* 2013;40:281. • Wang T, et al. *Science.* 2014;343:80. • Ran FA, et al. *Nat Protoc.* 2013;8:2281. • *Science.* 2013;342:1434.

Altering the way RNA forms proteins: Patel DJ. *Science.* 2014;346:542.

Precision medicine: Kandel ER. The new science of mind. *New York Times.* http://nyti.ms/15FfnE9. Accessed 8/25/14. • Aftimos PG, et al. *Discov Med.* 2014;17:81. • Kolata G. Finding clues in genes of "exceptional responders." *New York Times.* http://nyti.ms/1sh1Dc5. Accessed 10/9/14. • Doroshow J. In: Goldman L, Schafer AI, eds. *Goldman-Cecil Medicine;* 2016 (see bibliography). • Fact sheet: President Obama's precision medicine initiative. The White House. Office of the Press Secretary. http://wh.gov/iTBz3. Accessed 2/24/15.

Bibliography

American Psychiatric Association. *Diagnostic and Statistical Manual of Mental Disorders:* 5th ed. Washington, DC: American Psychiatric Association; 2013.

Barrett D. *Supernormal Stimuli : How Primal Urges Overran Their Evolutionary Purpose.* 1st ed. New York: W. W. Norton & Company; 2010.

Bloom RW, Dess NK, eds. *Evolutionary Psychology and Violence : A Primer for Policymakers and Public Policy Advocates.* Westport, CT: Praeger; 2003. See chapter:
Buss DM, Duntley JD. Homicide: An evolutionary psychological perspective and implications for public policy.

Boaz NT. *Evolving Health : The Origins of Illness and How the Modern World Is Making Us Sick.* New York: Wiley; 2002.

Brüne M. *Textbook of Evolutionary Psychiatry : The Origins of Psychopathology.* Oxford and New York: Oxford University Press; 2008.

Finlayson C. *The Humans Who Went Extinct : Why Neanderthals Died Out and We Survived.* Oxford and New York: Oxford University Press; 2009.

Gluckman PD, Beedle A, Hanson MA. *Principles of Evolutionary Medicine.* Oxford and New York: Oxford University Press; 2009.

Gluckman PD, Hanson MA. *Mismatch : Why Our World No Longer Fits Our Bodies.* Oxford and New York: Oxford University Press; 2006.

Goldman L, Schafer AI, eds. *Goldman-Cecil Medicine.* 25th ed. Philadelphia: Elsevier; 2016. See chapters:
Al-Awqati Q, Barasch J. Structure and function of the kidneys.
Anderson JL. ST segment elevation acute myocardial infarction and complications of myocardial infarction.
Appelbaum FR. The acute leukemias.

Bacon BR. Iron overload (hemochromatosis).

Bausch DG. Viral hemorrhagic fevers.

Blankson JN, Siliciano RF. Immunopathogenesis of human immunodeficiency virus infection.

Bosque PJ. Prion diseases.

Buchner DM. Physical activity.

Bunn HF. Approach to the anemias.

Crandall J, Shamoon H. Diabetes.

Doroshow J. Approach to the patient with cancer.

Dubose TD Jr., Santos RM. Vascular disorders of the kidney.

Garan H. Ventricular arrhythmias.

Ginsberg J. Peripheral venous disease.

Goldstein LB. Ischemic cerebrovascular disease.

Hansson GK, Hamsten A. Atherosclerosis, thrombosis, and vascular biology.

Jensen MD. Obesity.

Korf BR. Principles of genetics.

Lyness JM. Psychiatric disorders in medical practice.

Marks AR. Cardiac function and circulatory control.

Mason JB. Vitamins, trace minerals, and other micronutrients.

O'Connor CM, Rogers J. Heart failure: Pathophysiology and diagnosis.

Perlman A. Complementary and alternative medicine.

Quinn TC. Epidemiology and diagnosis of human immunodeficiency virus infection and acquired immunodeficiency syndrome.

Sawka MN, O'Connor FG. Disorders due to heat and cold.

Schafer AI. Approach to the patient with bleeding and thrombosis.

Schulman S, Hirsh J. Antithrombotic therapy.

Seifter JL. Potassium disorders.

Semenkovich CF. Disorders of lipid metabolism.

Slotki I, Skorecki K. Disorders of sodium and water homeostasis.

Steinberg MH. Sickle cell disease and other hemoglobinopathies.

Teirstein PS, Lytle BW. Interventional and surgical treatment of coronary artery disease.

Thakker, RV. The parathyroid glands, hypercalcemia, and hypocalcemia.

Victor RG. Arterial hypertension.

Weitz JI. Pulmonary embolism.

Grinde B. *Darwinian Happiness : Evolution as a Guide for Living and Understanding Human Behavior.* Princeton, NJ: Darwin Press; 2002.

Keedwell P. *How Sadness Survived : The Evolutionary Basis of Depression.* Oxford: Radcliffe Publishing; 2008.

Kurlansky M. *Salt : A World History.* London: Jonathan Cape; 2002.

McDougall C. *Born to Run : A Hidden Tribe, Superathletes, and the Greatest Race the World Has Never Seen.* 1st ed. New York: Alfred A. Knopf; 2009.

McGuire MT, Troisi A. *Darwinian Psychiatry.* New York: Oxford University Press; 1998.

Mielke JH, Konigsberg LW, Relethford J. *Human Biological Variation.* 2nd ed. New York: Oxford University Press; 2011.

Moalem S, Prince J. *Survival of the Sickest : A Medical Maverick Discovers Why We Need Disease.* 1st ed. New York: William Morrow; 2007.

Nesse RM, Williams GC. *Why We Get Sick : The New Science of Darwinian Medicine.* 1st ed. New York: Times Books; 1994.

Perlman RL. *Evolution and Medicine.* Oxford: Oxford University Press; 2013.

Pollan M. *The Omnivore's Dilemma : A Natural History of Four Meals.* New York: Penguin Press; 2006.

Relethford J. *Human Population Genetics.* Hoboken, NJ: Wiley-Blackwell; 2012.

Stearns SC, Koella JC, eds. *Evolution in Health and Disease.* 2nd ed. Oxford and New York: Oxford University Press; 2008. See chapters:

Ackermann M, Pletcher SD. Evolutionary biology as a foundation for studying aging and aging-related disease.

Leonard WR. Lifestyle, diet, and disease: Comparative perspectives on the determinants of chronic health risks.

Stringer C. *Lone Survivors : How We Came to Be the Only Humans on Earth.* 1st US ed. New York: Henry Holt; 2012.

Trevathan W, Smith EO, McKenna JJ, eds. *Evolutionary Medicine and Health : New Perspectives.* New York: Oxford University Press; 2008. See chapters:

Lieberman LS. Diabesity and Darwinian medicine : The evolution of an epidemic.

Wiley AS. Cow's milk consumption and health.

Wansink B. *Mindless Eating : Why We Eat More Than We Think.* New York: Bantam Books; 2006.

Zuk M. *Paleofantasy : What Evolution Really Tells Us about Sex, Diet, and How We Live.* New York: W. W. Norton & Company; 2013.

Index

About the Author

Dr. Lee Goldman is an internationally acclaimed cardiologist and dean of the faculties of health sciences and medicine at Columbia University, where he also serves as Harold and Margaret Hatch professor and chief executive of Columbia University Medical Center. He has been the lead editor of the last five editions of the renowned *Cecil Textbook of Medicine,* the longest continuously published text in the United States — which is now renamed *Goldman-Cecil Medicine* and which is sold worldwide in English and in six foreign translations. He is also the author of more than 450 scholarly research articles.

His work on determining which patients with chest pain require hospital admission (the Goldman Criteria) was featured in *Blink* by Malcolm Gladwell, who cited this work as a noteworthy example of how experts can make better decisions if they focus on a very limited number of truly important issues. He also is well known for his work predicting the cardiac risk of noncardiac surgery, including what is now widely known as the Goldman Index. Other articles have addressed limiting dietary salt intake, the adolescent obesity epidemic, and how to reduce the consumption of sugar-sweetened beverages.

Dr. Goldman has appeared on the *Today* show and has been interviewed on a wide range of topics related to cardiology and medicine. He frequently speaks at medical meetings as well as to lay audiences interested in health and medicine.

Dr. Goldman received his BA, MD, and MPH degrees from Yale University. He did his clinical training at the University of California in San Francisco (UCSF), Massachusetts General Hospital, and Yale–New Haven Hospital. Before joining Columbia, he was professor of medicine and of epidemiology at Harvard, and later was the chair of the department of medicine and associate dean for clinical affairs at UCSF.

Dr. Goldman is a member of the National Academy of Medicine, and he is the recipient of the highest awards from the Society of General Internal Medicine, the American College of Physicians, and the Association of Professors of Medicine.